印度标准化战略研究

浙江省标准化研究院
温州市标准化科学研究院 编著

浙江工商大学出版社
ZHEJIANG GONGSHANG UNIVERSITY PRESS
·杭州·

图书在版编目(CIP)数据

印度标准化战略研究 / 浙江省标准化研究院，温州市标准化科学研究院编著. — 杭州：浙江工商大学出版社，2021.11

ISBN 978-7-5178-4677-2

Ⅰ. ①印… Ⅱ. ①浙… ②温… Ⅲ. ①标准工作－研究－印度 Ⅳ. ①G307

中国版本图书馆 CIP 数据核字(2021)第 199101 号

印度标准化战略研究

YINDU BIAOZHUNHUA ZHANLUE YANJIU

浙江省标准化研究院　温州市标准化科学研究院　编著

责任编辑	王　琼
封面设计	沈　婷
责任印制	包建辉
出版发行	浙江工商大学出版社
	（杭州市教工路 198 号　邮政编码 310012）
	（E-mail：zjgsupress@163.com）
	（网址：http://www.zjgsupress.com）
	电话：0571-88904980,88831806（传真）
排　　版	杭州朝曦图文设计有限公司
印　　刷	浙江全能工艺美术印刷有限公司
开　　本	787mm×1092mm　1/16
印　　张	13.75
字　　数	237 千
版 印 次	2021 年 11 月第 1 版　2021 年 11 月第 1 次印刷
书　　号	ISBN 978-7-5178-4677-2
定　　价	48.00 元

前　言

印度是南亚次大陆最大的国家,面积约 298 万平方千米,位居世界第七,人口 13.9 亿(2021 年 8 月),是世界第二大人口国家,也是世界上民族较多的国家之一。1991 年开始的经济自由化使印度成为一个快速增长的主要经济体和新兴工业化国家。作为一个发展中的混合经济体,按国内生产总值计算,印度已经是世界第七大经济体,在市场汇率方面位居世界第六;按购买力平价计算,印度是第三大经济体。截至 2020 年,印度人均国内生产总值达到 1900 美元,排名 139 位,人均国内生产总值(以购买力平价计算)排名 119 位。2014 年以来,除 2017 年外,印度一直是世界上增长速度最快的主要经济体,已经超过了中国。在出口方面,印度已经成为软件业出口的霸主,也将成为全球金融、研究、技术服务等领域的重要出口国。投资方面,目前印度也已成为仅次于中国、美国之后全球第三大最具吸引力的投资目的地国。

国际标准化组织前主席海因茨(Heinz)先生曾经指出:"技术的进步和生产的发展与不同全球标准的发展有密切关系,因为后者为前者提供了通用性和互换性。"20 世纪 90 年代以来,技术经济与标准化的联系日益密切。印度在加入 WTO 之后,逐步重视工业标准化发展。在 2017 年 5 月举办的印度国家标准大会上,印度政府表示,正在起草一份综合性的国家标准化战略。2018 年 3 月 6 日,印度商工部在其主页上公布了印度最新国家标准化战略(2018—2023)(INSS)的草案,并面向国内外广泛征求意见。该文件是基于 2014—2017 年通过国家和地区标准化会议协商达成的广泛共识的结果,吸引了来自工会、州政府、行业企业、监管机构、国家和海外标准及合格评定机构、学术界和国际论坛的专家及利益相关方的广泛参与。2018 年 6 月 18 日,INSS 最终版面世。正如纳伦德拉·莫迪(Narendra Modi)先生在发表关于 INSS 信息时所言:"在我们人民的愿望和事业的推动下,一个充满信心和建设性地参与全球经济和贸易的新印度正在形成。值得注意的是,一个能够为经济带来深远的利益,并将提高印度制造品牌在全球市场的信誉的新的标准化框架正在被采用……我希望印度国家标准化战略能够利用全球范围内的最佳做法和创新,并确保为印度各部门的下一个转变提供便利的环境。"印度最新的标准化战略与政府提升"印度品牌"的重点一致,力求在"印度强健的质量生态系统"与协调、动态和成熟的标准框架之间取得平衡。

本书立足于印度最新标准化战略,总结印度标准化事务和进程的方向、框架与经验。从纵向来看,印度自独立以来的标准化实践与政策呈明显的阶段性发展态势,近期仍处于不断更新之中,其历史变化在一定程度上反映了印度经济和产业水平的变革;从横向来看,本书将聚焦最新一个阶段中,印度在制定和实施标准的过程中已形成的立法体系、管理架构、任务与措施等。在这些方面,印度当前的政策与实践构筑了一个复杂而独特的体系,提供了丰富的经验与成果,同时也存在着一定的缺陷和挑战。这些都将是本书重点关注的问题。

以此为基础,本书主要在两个方面有其关键意义。一是为印度标准化的研究构建基本框架。事实上,与印度标准化政策以及实践的丰富经验和成果呈鲜明对比的是国内对其研究的匮乏。在中国,目前尚无学术文献专门探讨印度标准化领域,即使是在其他议题的研究中提及这一问题的,也是罕见的。在印度,也仅有少量学术成果,从印度的标准机构参与 SPS/TBT 协定、印度标准局的运作、国际食品安全标准对于印度食品出口的阻碍、印度私有标准与竞争法的关系等特定方面进行分析。本书就印度的标准化政策与实践的框架与方向进行整体性的叙述,并对其中的关键部分和内容进行重点深入分析。二是为中国标准化事务的推进及中印两国之间标准化合作的开展提供思路。在中印政府之间,向来少有就标准化而进行的信息互通,这与印度政府对于标准化事务的日渐关注,以及标准化问题在中印间不断发展的贸易关系中权重的增加都极不相称。探索印度在这一领域的经验和政策方向,可以同时为中国的标准化工作提供有益启示与警示,而充分发挥标准化在推动双边贸易关系深入、良性发展中的作用,也需要以印度的标准化政策和实践为基础,探寻两国间可行的对话与合作机制。

目　　录

第 1 章

印度标准化战略的发展历程

随着经济的快速增长、发展规模的不断扩大,以及在全球贸易中新兴地位的持续提升,印度在和谐、有活力和成熟的标准框架下建立一个强大的"质量生态系统"变得至关重要。在过去的 70 年中,印度已经制定了 20000 多个标准,其中 50％为产品标准,其余则为测试方法、术语、操作规范等支撑标准。对于新标准的提案或国际标准的采用均由印度标准局的各理事会、技术委员会和标准开发组织(Standard Development Organization,SDOs)决定。然而,正如印度最新标准化战略文件所强调的,在印度,海外机构制定的标准正在被广泛使用,却没有得到采纳或加以调整。另外,在没有使用标准的地方也有很大的差距,特别是在服务部门,有些领域则只有指导标准或规范存在,却没有相关的产品标准。因此,印度政府迫切需要将标准化作为所有部门的关键领域,通过对话和建立规范的流程,以明确标准开发的需求,并将其列为优先级任务。综观印度标准化机制,可以看出其长期的演变和发展的过程呈现出明显的阶段性特征,且每个阶段与特定时代下印度政府的宏观经济政策紧密地关联在一起。

1.1 印度标准化战略的阶段性特征

在 20 世纪上半叶,印度尚属英国殖民地时,印度的一些部门已开始了标准化工作。例如,《农业生产(分级与标志)法案 1937》[*Agricultural Produce (Grading and Marking) Act*,1937]、《药品与化妆品法案 1940》(*Drugs and Cosmetics Act*,1940)等法律法规文件都是在殖民地时期制定的,其中涵盖农业或药品这些专门产业领域的标准问题,且它们的法律效力延续至今。不过这一时期,印度的标准化管理呈现出较零散的状态,未形成覆盖面较广的管理架构,且基本上只限定于政府的采购行为。印度政府有意建立覆盖面较广、成体系的标准化管理架构,是自印度独立开始的。

自 1947 年印度独立以来,印度的标准化战略进程呈现出明显的阶段性特征。以关于标准化的专门法案的颁布,以及在中央政府层级成立一个统一处理标准化事务的机构为标志,印度标准化战略迄今为止的发展历程可以划分为两个阶段,第二个阶段中又包含了三个小的阶段。之所以以上述事件为标准化进程中的阶段性标志,是因为专门法案的颁布及中央政府中标准化管理机构的成立,是印度政府决心整合分散于各产业部门的标准化事务,在全国范围内建立起统一的标准化法律与行政管理架构的最主要象征。

  具体而言,从 1947 年印度独立开始至 1985 年,是印度标准化战略进程中的第一个阶段。伴随着印度独立,印度政府在 1947 年成立了印度标准协会(Indian Standards Institute,ISI),标准化因而成了当时的印度政府构建后殖民地时期经济管理架构中的一项工作。拉尔·魏尔曼(Lal Verman)成为印度标准协会的第一任会长,在启动印度标准化进程中发挥了领导性的作用。在接下来的几年内,印度政府进一步完善标准化的管理,主要行动包括在 1952 年颁布《印度标准协会(认证标志)法案 1952》[*Indian Standards Institute*(*Certification Marks*)*Act*,1952],并以此法案为依据,在 1955 年启动认证标志计划。通过这一全国性的认证项目,印度标准协会为符合标准并进行了认证的产品贴上"ISI 标志"。

  自《印度标准局法案 1986》(*The Bureau of Indian Standards Act*,1986)颁布至今,是印度标准化战略进程中的第二个阶段。与前一阶段相比,这一阶段对有关标准化的国家专门法案以及中央政府机构都进行了彻底更新。这也显示出印度政府加强了对标准化工作的重视程度,标准化管理的效率、执行力、体系化都得到了大幅度提升。

  尽管 1947 年印度标准协会的成立标志着印度政府开始在国家层面上涉足标准化管理,但该协会的职能及其实行的标准化项目的效果都很有限。印度标准协会的主要工作只限定于标准认证,而不参与标准制定及其他与标准相关的诸多工作;在市场上,参与标准认证并贴上"ISI 标志"的产品非常稀少,印度标准协会的产品认证项目在印度市场上所起到的作用十分有限。为了改变这一状况,加强标准化在印度产业发展中的地位和作用,印度政府于 1986 年颁布《印度标准局法案 1986》,并根据此法案于次年正式成立印度标准局(Bureau of Indian Standards,BIS)。该法案确立了印度标准局在印度标准化进程中的核心地位,并规定了印度标准局在标准制定、产品认证等一系列标准化事务上所拥有的权力。随着印度标准局的成立,印度标准协会被撤销,其资产、人力等全部并入印度标准局,而后者在规模、管理体系设置等方面都在印度标准协会的基础上跃进了一个层级。

  自莫迪政府执政以来,标准化受到进一步的重视。印度政府于 2014 年开始启动"国家标准大会"(National Standards Conclave)。2015 年,印度政府开始修订《印度标准局法案 1986》,印度上院(Rajya Sabha)于 2016 年 3 月 8 日通过了《印度标准局法令

2015》(*Bureau of Indian Standards Bill*, 2015)，对前一部法案做了大幅度的修改，后在此基础上通过了《印度标准局法案 2016》(*Bureau of Indian Standards Act*, 2016)。在经过一段时间的公示之后，印度政府于 2017 年 10 月 12 日宣布该法案正式生效。《印度标准局法案 2016》的生效自动取代了 1986 年的法案，以此为法律依据，印度标准局继续担当印度标准化管理的核心机构，且其职权得到进一步强化。因此，新法案的颁布和生效可谓在印度标准化进程的第二个大阶段之中开启了一个新的发展时期，该法案的执行效果则有待继续观察。

2018 年 1 月，莫迪总理在达沃斯论坛开幕致辞上大谈"印度梦"时首次提出了"到 2025 年，印度将成为一个 5 万亿美元的经济体"的目标；同年 3 月，在一次小组会的讨论中，印度经济事务秘书长 Subhash Chandra Garg 也宣称，"我认为到 2025 年我们实现 5 万亿美元的经济体目标是非常可行的，而其实这也是一个合理的目标"。不管实现这个 5 万亿美元的目标难度如何，印度经济的快速增长已经是事实，加上发展的规模及其在全球贸易中出现的相关性，使得印度建立一个协调、动态和成熟的标准生态系统和强大的"质量基础设施"迫在眉睫。于是，在 2018 年 6 月 19—20 日，印度商务部和印度工业联合会(CII)在新德里举行的第五届国家标准会议上，印度商务和工业部部长苏雷什·普拉布(Suresh Prabhu)首次发布了印度史上第一份综合性的国家标准化战略，简称"INSS 2018—2023"(Indian National Strategy for Standardization 2018—2023)。这份战略的形成并非一时兴起的结果，而是在 2014—2017 年间通过多次的国家和地区标准会议，吸引了来自联邦和州政府、行业企业、监管机构、国家和海外标准及合格评定机构、学术界和国际论坛的专家和利益相关方的广泛参与。该战略为印度提供了一个愿景，使其在商品和服务的生产和分销方面达到最高质量标准，以期重振印度品牌。INSS 报告阐述了质量生态系统的四大支柱：标准制定、合格评定和认可、技术法规和 SPS 措施、认识和教育。

1.2　印度标准化战略与宏观经济进程的关联

上述印度标准化战略的阶段性进程与该国自独立以来的宏观经济进程之间存在着密切的关联。在不同的宏观经济政策时期，印度政府对于标准化的需求的方向和程度也有所不同,这是其标准化管理事务呈现阶段性进展的重要原因所在。自独立以来,印度中央政府的经济政策及相应的印度的经济模式,主要有两个阶段。从尼赫鲁时期(1947—1965)到 20 世纪 80 年代,印度政府实行高度集中的、国有制的计划经济政策,并以进口替代模式为主,经济发展相对封闭,独立于世界经济体系之外。这一经济政策和发展模式自 20 世纪 80 年代初开始有了小幅度的调整,1984 年上台的拉·甘地接续前一届英·甘地政府的步伐进一步推进改革,在一定程度上引入市场竞争。在此基础上,到了 20 世纪 90 年代初,印度经济发展的方向发生骤然转变,最主要的是开启了对外经济自由化的进程,进口替代型模式向出口导向型转变,外国商品和资本开始越来越多地进入印度市场,国内的国有制经济模式也开始了其私有化进程。自此开始,印度每一届中央政府的经济政策路线基本上都保持一致,以大力推行对外经济开放等为己任,尤其是当前的莫迪政府,极力推进土地、劳工、税收等方面的经济改革,创造吸引外来投资的环境,并以"印度制造"为旗号,大力推进印度产业水平的发展及其出口。不过值得注意的是,即使是在印度政府开始走对外经济自由化路线之后,印度经济政策也并非完全摆脱了贸易保护的取向,而是将开放与保护微妙地结合起来,成了当前印度经济政策及经济发展的一个特质。

印度标准化战略的阶段性发展可以从上述宏观经济政策和模式的变化中得到解释。尼赫鲁政府时期大力推行以重工业为重点的工业化进程,并强调中央政府对工业化的计划与把控,在这一框架之下,需要对工业发展实施许可证法,对工业的生产质量和商品流通进行控制。正是在这一方面,标准化的作用得到了关注。而拉·甘地的经济政策改革包括了工业发展重点的调整,即由以往的重工业向消费品工业和电子工业等领域倾斜,倡导将竞争机制更多地引入工业生产,减少对本国工业的政策性保护。自此开始,无论是消费品等领域产业的发展及消费者对质量的需求,还是私有化和对外市场开放带来的市场竞争以及促进产品出口的需求,都要求推行一个新的质量管理体系。

可以认为,《印度标准局法案 1986》的颁布以及印度标准局的运作,是印度政府在新的宏观经济目标之下对标准化认知和需求的更新。从印度标准化进程的第一阶段到第二阶段,不仅标准化得到的重视程度有所加强,还意味着标准化开始在新的层面上对印度经济的发展发挥作用,即不再是中央政府加强中央集权经济控制的工具,而是成了印度政府维护市场竞争秩序、提升经济开放情况下的产业竞争力、促进印度经济融入世界经济体系的重要而有效的途径。不过,在新的阶段,由于经济开放性和保守性的并存,标准化实际上被印度政府同时用以实现其经济开放和贸易保护这两种需求,只是此时贸易保护的需求是整体上经济开放背景下的自我保护,已与尼赫鲁时期以标准化服务于国家计划经济的状况有了本质上的不同。

第 2 章

印度标准化战略的配套立法

　　1986 年,印度政府颁布了《印度标准局法案 1986》,大幅度地改进了印度的标准化管理机制,立法在印度标准化战略中的重要性可见一斑。围绕着标准化问题,多部专门法案或对此领域有所涉及的法案构成了一个复杂的立法体系。本章将首先概述各类法案如何构成一个印度标准化战略的立法体系,继而对其中若干处理标准化问题的专门法案内容进行详述。

2.1　印度标准化战略的立法体系

　　《印度标准局法案 1986》及《印度标准局法案 2016》是印度政府先后搭建印度标准化立法框架的重要尝试,但无论是在 1986 年之前还是以后,都另有多部法案或多或少地从不同角度涉及标准化问题。

　　从纵向上来看,印度标准化战略的立法体系中不仅不断出现新的法案,而且前后不少法案之间也存在着新旧更替的情况(见图 2-1)。如《印度标准局法案 1986》颁布之后,《印度标准协会(认证标志)法案》被废,而《印度标准局法案 2016》自 2017 年 10 月 12 日开始生效则意味着 1986 年法案自动被替代。在食品安全标准领域内,印度一度存在着繁多的相关法案,如《防止伪劣食品法案 1954》等,而《食品安全与标准法案 2006》(*Food Safety and Standards Act*,2006)的出现则替代了该领域内的诸多法案,构建了食品安全标准的新的统一法案。

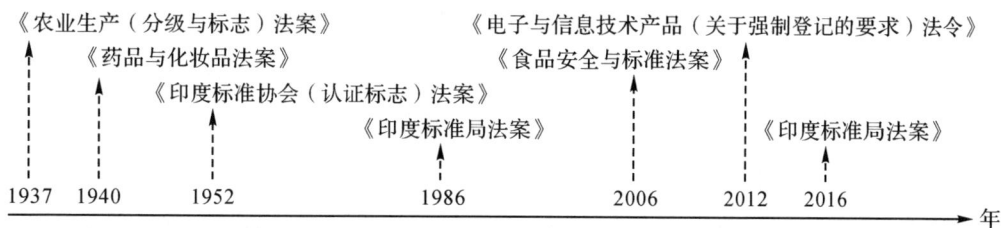

图 2-1　印度代表性的标准化法案的时间轴

　　就当前正在生效的法律而言,构成印度标准化立法体系的主要包括聚焦标准化问题的专门法案和部分涉及标准化问题的其他法案。

　　在专门法案方面,主要包含如下 4 种类型的法律文件。

　　其一是为了赋予专门的标准化机构以法律地位,并规定其基本的运作方式、权限和责任而颁布的法案。当前,这样的法案有两部,即《食品安全与标准法案 2006》与《印度标准局法案 2016》,它们分别为印度食品安全与标准局(Food Safety and Standards Authority of India,FSSAI)和印度标准局的运作奠定了法律基础。而以规定管理机构的运作为途径,这两部法案实际上是为印度标准化事务的统一管理机制构建起了法律框架。只是《印度标准局法案 2016》并不针对特定领域,而《食品安全与标准法案 2006》则聚焦食品安全领域内的标准化管理。

其二,在上述两部法案的框架下,分别存在着若干具体的法律规定,继续处理这两部法案管辖范围内的一些细节事务。《印度标准局法案 1986》中说明,经中央政府允许,印度标准局的执行委员会可以指定与该法案不相违背的规定与规则,以进一步执行这部法案。基于此,在 1986 年法案颁布之后,又陆续出台了多个规定,对印度标准局的一些具体事务进行规定,它们同样享有法律地位。这包括《印度标准局规则 1987》《印度标准局(咨询委员会)规定 1987》《印度标准局(认证)规定 1988》等。在 2016 年法案生效之后,这些规定将在不违背新法案的情况下继续运作。在《食品安全与标准法案 2006》的框架下,同样存在着数项规定。与印度标准局法案框架下的规定相比较,这些食品安全领域内的规定更为关注对食品安全具体标准的设定及其执行方式。其中,《食品安全与标准(食品标准与食品添加剂)规定 2011》《食品安全与标准(污染物、毒素与残留物)规定 2011》及《食品安全与标准(食品或保健品、营养品、特殊膳食品、特殊药用食品、功能食品及新型食品)规定 2016》对多种食品的具体技术标准做出了详尽的规定;《食品安全与标准(食品经营的许可与登记)规定 2011》《食品安全与标准(销售的禁止与限制)规定 2011》《食品安全与标准(包装与标签)规定 2011》《食品安全与标准(实验室与样品分析)规定 2011》《食品安全与标准(食品召回程序)规定 2017》《食品安全与标准(进口)规定 2017》《食品安全与标准(非特定食品与食品配料的批准)规定 2017》则对食品标准执行的不同环节做出了规定。

其三,对标准化管理中的部分具体环节做出规定,但不属于前述法律框架的法案。这主要包括《重量与计量标准法案 1976》(*Standards of Weights and Measures Act*,1976)、《出口(质量管控与检查)法案 1963》[*Export (Quality Control and Inspection) Act*,1963]及其 1984 年的修正案。前者对经包装的产品在重量、计量和数量上的标准及标示这些内容的方式做出了具体的规定,后者则规定了印度出口过程中应用标准化以进行出口产品质量管控的具体方式。

其四，对需纳入强制认证范畴的产品做出法律规定的法案。在《印度标准局规则 1987》中有这样的说明，"印度标准是自愿并向公众开放的，其实施由相关方自行决定。然而，在这样的情况下，印度标准将具有约束力：在合同中得到规定、法律中有所提及，或政府的具体行政令规定了强制认证"。与之相对应的是，印度中央政府陆续颁布了多个法令，每个法令针对某一项或某一类具体产品，对其被纳入强制认证范畴的情况做出规定。每出台一个这样的法令即意味着在印度标准局制定的"印度标准"（Indian Standard）的框架下，强制认证的范畴中就增添了一种或一类产品。例如，有 13 种水泥产品属于强制认证的范畴，这由《水泥（质量控制）法令 2003》[Cement（Quality Control）Order, 2003] 所规定；16 种需要进行强制认证的家电产品由《电线、电缆、电气及保护装置与配件（质量控制）法令 2003》[Electrical Wires, Cables, Appliances and Protection Devices and Accessories（Quality Control）Order, 2003] 所规定；《储气罐法规 2016》（Gas Cylinder Rules, 2016）则规定了多个需强制认证的气缸、阀门及调压器产品；《钢铁及钢铁制品（质量控制）法令 2012》[Steel and Steel Products（Quality Control）Order, 2012] 中涉及 35 种钢铁及钢铁制品，将其纳入强制认证的范畴；《电子与信息技术产品（关于强制登记的要求）法令 2012》[Electronics & Information Technology Goods（Requirements for Compulsory Registration）Order, 2012] 则规定了电子与信息技术产品进行强制认证的范围；等等。

除此以外，还有相当多并非专门聚焦标准化问题的法案，其中包含着一定的有关标准化的内容。这主要是一些聚焦具体的产业领域的法案，其中涉及对该产业领域中标准化运作方式的规定。早在殖民地时期，《农业生产（分级与标志）法案 1937》及《药品与化妆品法案 1940》即分别规定了农业分级标准以及药品与化妆品的标准。例如后者意在保障销售于印度市场的药品和化妆品安全、有效、符合质量标准要求，其中详细规定了各类药品与化妆品各自需要符合的基本标准，以及药品不符合质量标准时应采取的检测和分析措施，等等。印度独立后颁布的诸多类似法案中也都包含有标准化的内容。例如，《印度国家建筑法规》最初颁布于 1970 年，后经 6 次修订，形成了《印度国家建筑法规 2016》（National Building Code of India, 2016），为全国范围内建筑建设活动的管理提供了纲领。该系列的法规包含了对于建筑设计和建设的诸多要求，包括防火、材料、结构设计等方面的标准。为了适应当前不断扩大的建筑活动，最新版本的法规更是增加和更新了有关建筑的结构性安全、无障碍环境、节能环保等方面的诸多要求。另有一些并非针对特定产业领域的法案同样关涉标准化问题，它们主要涉及标准化管理运作的一些具体环节。

2.2 主要标准化法案的内容

2.2.1 《印度标准局法案 1986》

在《印度标准局法案 1986》颁布以前，印度标准协会的成立以《社团注册法案 1860》（*Societies Registration Act*, 1860）为法律依据，而《印度标准协会（认证标志）法案 1952》则仅涉及有关标准认证的相关事项。相比之下，《印度标准局法案 1986》则为中央层面的标准化部门的运作及其管理之下的一系列标准化活动提供了较为全面的法律依据。

在内容结构上，《印度标准局法案 1986》既包括了对印度标准局的组建、人事架构及财务等方面的日常行政运作的说明，也包含了对印度标准局的权力与职能、标准标志的使用规范、认证许可的颁发等相关事项的规定。尤其是后一方面，直接涉及如何通过印度标准局框架下的标准制定、产品认证以及颁发许可等在全国范围内管理标准化活动。

具体而言，在标准制定方面，印度标准局有权制定和发布"印度标准"，或将其他机构所制定的标准认定为"印度标准"。在产品认证方面，对于"印度标准局认证标志"（Bureau of Indian Standards Certification Mark）的授予、更新、暂停或撤销都在印度标准局的职权范围之内。不仅如此，《印度标准局法案 1986》还规定了印度标准局有权在必要的时候提取样品，检查使用了上述标志的产品是否符合"印度标准"，以及在使用上是否拥有合法的许可，等等。更进一步地，为了避免对于"印度标准"和认证标志的滥用与误用的情况出现，法案对一些相关情况做出了禁止性规定，如：在未获许可的情况下，任何人不得在任何产品、流程、专利或商标中使用印度标准局的认证标志或相类似的仿制标志；在未经标准局提前允许的情况下，任何人不得在任何标志或商标中包含"印度标准""印度标准规格"或其他类似表达。在许可颁发方面，印度标准局的职权包含了对相关认证许可的授予、更新、暂停或撤销，授予和更新许可的情况及费用由标准局的相关规定决定。法案同时规定了对有关许可的决定不满的个人向中央政府提起申诉的情况,包括相关流程及申诉在何种状况下不会被接受等。

2.2.2 《印度标准局法案 2016》

如前所述,《印度标准局法案 2016》于 2017 年 10 月 12 日正式生效。自此开始,《印度标准局法案 1986》自动失效,印度标准局的运作及标准化管理的一系列事项开始以 2016 年法案为依据。在内容上,最新的这一部法案既是对 1986 年法案的总体框架的继承,也在不少具体的条款上做出了调整、更新和增添。继承的方面反映了自 1986 年以来构建的以印度标准局为核心的全国性标准化管理方式和途径未有改变,这主要表现为两部法案在整体的内容结构框架上基本一致,有关印度标准局的人事架构、财务等日常行政运营的法律说明也只有些微调整。更新的内容则反映了印度政府加强印度标准局在国家标准化管理体系中的作用和地位,强化对于标准化活动的管理的决心。

相较于 1986 年法案,《印度标准局法案 2016》的内容调整主要体现在印度标准局的权力与职能、产品认证和许可颁发等方面。具体而言主要有以下几个方面。

其一,国家标准化管理覆盖的范围得到了扩大。1986 年法案仅仅说明是有关"商品的标准化、标记及质量认证",新的法案则在开篇即表明其所涵盖的有"商品、物件、流程、系统和服务的标准化、合格评定和质量保证",反映了印度政府对于系统和服务等超越传统产品标准化管理领域的重视逐渐加强。

其二,更新了有关强制认证的规定,使强制认证在印度标准化管理中占据更为重要的地位。1986 年法案的第 14 条涉及强制认证的问题:在咨询印度标准局之后,当中央政府认为为了公众利益而有必要的时候,可以发布政令以通知任何产业中的任何产品或流程必须要遵从"印度标准",并要求强制进行产品认证,使用标准标志。而在 2016 年法案中,中央政府可将产品、流程、系统或服务纳入强制认证范畴的必要情况则不仅仅是前述的公共利益需求这一笼统表述,还更加广泛也更深入细致地包含了另外 4 种情况,包括保护人类、动植物健康的需求,维护环境安全的需求,防止不公平贸易发生的需求,以及国家安全的需求,即当印度中央政府在这些情况下认为有必要时,都可扩大强制认证的范畴。此外,2016 年法案还特别规定了对贵金属制品进行强制认证的情况,即在咨询印度标准局后,中央政府有权在认为有必要的时候,规定贵金属制品必须在获得纯度认证之后方可在市场上流通,甚至是只能通过经认证的、符合相关规定的销售网点进行销售。上述几条规定,反映了印度政府对于实施强制认证的必要性的聚焦,并认为有更多的产品和服务应当被纳入强制认证的范畴,从而实现其在贸易、公共安全等多个层面的目标。

其三,印度标准局的职权在较大程度上得以加强。有关印度标准局的职权,1986年法案基本上只是规定了其基本方面,如制定和发布"印度标准"、进行产品认证、许可授权等。在此基础上,2016年法案对印度标准局在标准标志的使用等方面的职权做出了诸多详细规定,并赋予了该局一定的强制性权力。这主要有:印度标准局可对标准标志及商标的要求做出更详细的规定;可对任何商品或服务开展市场调查,可对任何带有标准标志的产品或服务提取样品、进行检查,以确定其是否符合相关标准的要求,或标准标志是否在拥有许可的情况下被正确使用;对于未通过上述检查的产品或服务,印度标准局有权要求相关制造商或商家根据情况采取相应措施,包括停止供应和售卖,召回市场上不符合规定的商品,修复或替换相关商品,向消费者支付赔偿金,并在造成伤害的情况下承担相关法律责任。通过这些规定,印度标准局就标准认证和使用进行市场监管方面获取了更多、更具法律强制性的权力,同时,也增加了不符合法律规定而滥用或误用标准标志的相关方所需要付出的法律代价的规定,这为标准化领域市场秩序的整顿奠定了重要基础。

其四,对标准的制定、产品认证、许可获取等方面做出了更为严格和细致的禁止性规定。如前所述,相关禁止性规定在1986年法案中已经出现,在此基础上,2016年法案中有更多的条款和内容涉及这一方面的规定,如:未经印度标准局授权,任何人不得制定、发行、更新"印度标准",或印度标准局的其他出版物;对于被纳入前述强制认证范畴的产品,未经印度标准局许可,任何人不得进口、售卖、以商业利益为目的进行储存或展览,任何人不得制造、进口、售卖未获得认证的产品;未经印度标准局授权,任何检测及认证机构不得使用标准标志。尤其值得重视的是其中对于"进口"的涉及,这一词汇在1986年法案中未曾出现。这意味着对进口市场进行更为严格的把控。

上述几点法案内容上的变动,实质上标志着印度政府的标准化战略的未来发展方向,即增加标准化管理的覆盖面、提升标准局的职权和地位、强化以标准化为途径而对印度市场展开的强制管理等。

第 3 章

印度标准化战略的目标定位

印度政府发布的第 12 个五年计划(2012—2017)明确提出,标准机制是印度工业界需要学习以建立良好的产业生态系统的重要方面之一,"它们(标准)使企业,尤其是小型企业具备知识基础,并且是在国内外削减企业与消费者、供应商之间的交易成本的方式之一"。为发挥标准化在国家经济发展中的作用,印度政府为标准化工作设定了短期和中长期等不同层次的目标。2018 年 6 月,印度首部综合性标准化战略 INSS 2018—2023 的问世从战略上完成了短期目标的任务,该战略针对"标准开发""合格评定、认可和计量""技术法规和 SPS 措施""意识和教育"四大主体内容设定了详细的目标,虽然一些目标和相关活动可以在较短时间内完成,但战略中所涉及的所有要素将在 5 年时间(2018—2023)内进行和完成。本书将印度的最新标准化战略定位于短期目标,并进行讲解。然而值得注意的是,印度的短期目标和中长期目标之间,既有相互协调之处,也存在着一定的矛盾。如何在标准化事务的进程中将两者结合起来,正是印度标准化的问题所在。

3.1　印度标准化的短期目标定位

其一,实行适应本国产业发展水平的标准化措施。从总体上来看,印度国内的产业,尤其是第一产业和第二产业,国际竞争力较低,标准化成为印度政府借以减小国内部分弱势产业受外来冲击的重要方式。在多种场合,印度政府强调在国内设定一定的标准并据此对进口进行管控,意在将低质量的产品拒绝在印度国门之外,从而保障国内消费者等群体的安全。例如,在第四届印度国家标准大会期间,印度政府发布的相关声明称,如果印度缺乏足够的技术规定,其国内消费者、本土产业甚至是整个经济将因不合标准、不安全的产品的进口而蒙受损失。实际上,对本土脆弱产业进行保护更是印度政府借助标准化措施控制进口的核心目的。印度第 12 个五年计划中对此有着非常明确的表述:"技术标准的缺乏使得低质量的产品能够非常容易地进入印度市场。这损害了本土产业,因为本土产业无力在价格上与这些进口产品进行竞争。……这样的例子在移动电话、移动电话的电池、电子血压测量仪、装饰灯、医疗器械等方面比比皆是。"根据相关报道,更是有印度官员直截了当地表示:"在颁行任何技术规定之前,我们需要对贸易赤字进行分析,因为在相当一些领域中,我们的国内生产较为薄弱。"可以发现,印度近年来越来越密集地颁布相关规定,将更多的产品纳入强制标准认证的范畴,并使其适用于产品进口,而这些产品如钢铁、电子产品等,主要集中在印度政府予以发展和保护的一些重点产业领域。后文将对强制认证进行更为详细的论述。

其二,制定明确的国家标准化战略。长期以来,印度政府尽管在推进标准化工作方面做出了诸多努力,但直到 2018 年新战略问世之前始终未明确提出国家标准化战略。印度政府逐渐意识到,要推动标准化的深入有序发展,给予标准化工作足够的国家层面的认可与重视,在经济增长和产业发展中恰当定位标准化的位置,就必须制定和发布国家战略。在印度标准局列出的组织目标中,已包括了发展国家战略这一项。2017 年,国家标准化战略的制定已被印度政府提上议事日程,以国家标准大会为平台,印度政府召集来自公私部门的众多利益相关者,就制定标准化战略、准备"印度国家标准化战略"文本进行讨论。2018 年的最新战略 INSS 2018—2023 充分考虑了印度各部门的发展现状、现有的质量基础设施、国内经济发展以及与商品和服务贸易有关的政策方向,提出了质量生态系统的四大支柱,即标准开发,合格评定、认可和计量,技术法规和 SPS 措施,意识和教育。INSS 为印度的政治和行政领导层在如何最优化利用标准化、技术法规、质量基础设施以及促进印度在全球经济中的利益和福祉等相关活动中提供指导。其战略性意图主要基于:

(1)定位标准是所有与货物和服务有关的经济活动的关键驱动力。

(2)在考虑利益多样化和专门知识可利用化的基础上,开发一个制定标准的综合生态系统。

(3)利用标准为国内工业提供公平的竞争环境,提高印度商品和服务在国内和国际市场的竞争力。

(4)采用标准化、合格评定和认证以及技术法规方面的最佳做法,并为其有效的管理创建集成的基础设施、路线图和机构。

(5)在相关领域的顶尖国际论坛中发挥积极作用并担任领导职务。

(6)建立影响印度商品和服务市场准入的标准、技术法规以及合格评定全球发展的应对机制。

(7)将战略和其他与贸易、工业、消费者以及环境有关的国家政策相统一。

INSS 2018—2023 的四大主体内容中,"标准开发"设定了"融合印度所有标准开发活动"等八大目标,"合格评定、认可和计量"包含了"提高合格评定方案在国内外市场的可信度"等五大目标,"技术法规和 SPS 措施"则设立了"建立对高效监管实践和监管影响评估的良好理解"等七大目标,"意识和教育"包含了"提高利益相关者的意识"等三大目标。主要内容如下。

3.1.1　标准开发

INSS 关于"标准开发"的愿景是为经济增长和经济引领建立国家标准文化,其主要使命是发展一个动态的、协调的和优先驱动的标准生态系统,推动跨部门的发展,提高印度产品和服务的竞争力,并培养印度在标准化方面的全球领先地位。"标准开发"包含了八大目标。

目标 1:融合印度所有的标准开发活动。包括:①增强标准开发组织(SDOs)的能力,加快标准制定,适应技术发展的步伐;②鼓励在具有国际联系的新兴技术领域中建立新的 SDO;③必要时采用 SDO 标准作为国家标准;④设立 SDOs 识别体系;⑤避免重复、冲突和重叠;⑥确保所有市场相关性标准的及时开发。

自 1947 年以来,印度的标准制定过程由印度标准协会主导。如前文所介绍的,在有限的范围内,特定部门的标准化工作由 25 个以上的其他机构进行,包括各部、监管机构、公共事业部门、技术开发机构、商品委员会、行业和专业团体等。近年来,越来越多的 SDOs 在印度设立了办事处,以便聘请专家并使用其标准支持产业发展。

虽然 BIS 通过代表部门利益的 14 个理事会下的委员会来开发印度标准,但基于标准开发的国际实践守则,其他 SDOs 基本上依赖专业的知识进行标准编写,并遵循自由的程序。目前为止尚未存在一种可识别或整合这些标准作为国家标准的系统。

为了扩大基础和加快标准制定活动的步伐,有必要增强现有 SDOs 的能力和资源基础,并鼓励在新兴地区和简短技术中建立新的 SDOs,尤其是数字技术、可持续实践、清洁能源和智能城市。这些 SDOs 必须与相应的国际机构建立联系并做出工作安排,以确保印度工业在标准的可用性方面不存在差距。

根据《印度标准局法案 2016》的授权,印度标准局仍是最高的国家标准机构,将继续监督其自身框架下标准化活动的协调发展以及与其他 SDOs 的谅解备忘录,其中包括在国家需要和优先事项要求时,采用其标准作为国家标准。这样的安排将确保多个机构的标准化活动不会重复或重叠。它们应始终负责确保有效的标准始终适用,并通过适当的市场调查、环境扫描和与全球范围内制定标准的比较研究来反映技术和工业实践状态。

应建立认可机制，以确保它们执行 WTO 的业务守则、标准制定原则以及有关委员会在这方面的任何相关决定。

目标 2：建立新的标准识别与制（修）订动态机制。包括：①使标准化成为所有行业的一个重要优先领域；②建立对话论坛和进程，阐明并优先考虑标准发展的需要；③通过标准为印度企业创造机会；④关注与经济、社会和可持续发展相关的关键部门。

在过去的 70 年中，印度已经开发了 20000 多项标准，其中将近 50% 是产品标准，其余的是包括测试方法、术语、操作规范等在内的支撑标准。新标准立项或国际标准的采纳由 BIS 的各分部理事会和技术委员会决定，而其他 SDOs 则通过委员会或行政决定做出类似决定。

印度国内目前的系统未能提供充分的机会或渠道来满足所有潜在标准用户的需求。因此，许多海外机构开发的标准在印度被广泛使用，但没有被等效采用或修改采用。目前特别是在服务部门，尚没有可供使用的标准，在有些领域，仅存在指导性标准，但不存在相关的产品标准。因此，印度迫切需要通过开展讨论会和制定流程，明确并优先考虑不同部门的标准开发需求。满足需求的最佳候选者则是其下属的部委和政策论坛、相关的行业机构、出口促进机构和商品委员会等。这就需要在每个机构中建立一个标准论坛，负责与企业、专业机构建立利益相关方磋商和对话论坛，以确定差距、整理需求，并就所需标准的及时开发或修订与 BIS 及相关 SDOs 进行协调。

利益相关者协商期间的主要考虑因素是技术的识别，市场、行业和州级要求，在这些方面，标准化工作可以为印度企业创造机会。

目标 3：包括各邦和中小微企业在内的所有利益相关者都参与标准化的开发。包括：①提高企业、政府和包括消费者在内的民间社会团体对标准和合格评定实践的作用及益处的认识；②为标准相关活动创建国家和地区论坛；③建立筹资机制，对各利益相关者的参与予以支援。

有效的标准制定需要利益相关方和专家的积极持续参与。为了实现这一目标，必须使其对 BIS 和其他 SDOs 正在开展的标准计划产生广泛的兴趣以及对标准制定的意识，并在资金有限的情况下吸引资金投入，特别是对于中小微企业以及包括消费者团体和专家在内的民间社会组织。

自 2014 年以来启动的标准会议已成为一个良好的论坛，可以为标准化、合规性评估和技术法规工作提供信息和教育培训。后期仍需继续在中央和州一级轮流举行更多的标准会议。

随着时间的推移,标准制定过程应成为一项分层活动,可以邀请各州和各地区标准论坛提供意见。标准委员会应根据工业或经济活动的强度,派代表直接参与此类论坛。

中央政府为参与标准制定活动提供广泛、自由的资金援助,以资助中小微企业、专家、非营利机构和民间社会团体参与国家和国际标准委员会的工作。必须鼓励州政府资助并参与其感兴趣领域的国家标准制定工作。

目标 4:使标准与国际标准协调一致。包括:①尽可能使国家标准与国际标准协调一致;②避免国家标准与市场驱动标准不协调;③特别关注并汇集融合技术方面的专业知识。

必须使印度标准与减少贸易技术壁垒、改善印度产品及服务的市场准入的国际标准相协调。为了确保印度企业在国内和海外市场上的竞争力,商品和服务必须符合全球公认的标准,也必须符合国家的优先事项和要求。虽然应适当考虑国家因素,但这些要求应是最低限度和最不严重的贸易限制。BIS 和其他印度标准组织、参与国际标准组织[ISO、IEC、ITU、食品法典委员会、世界动物卫生组织(OIE),以及国际植物保护公约(IPCC)]需要系统采用、适应和调整相关标准与相应的国际标准。该工作的重点应放在产品、设备、硬件标准上,同时采用、适应相关规范、测试方法和协议标准。

除了国际标准组织制定的标准之外,由若干专业或私立机构制定的标准也被认为具有重要的市场地位和广泛的用户群。BIS 和其他从事各自领域的 SDOs 需要确保其标准不会与产业、企业产生冲突,并努力避免与私营标准不一致,特别是在需要强制符合国家标准的情况下。

信息通信技术和数字技术在制造业和服务等领域的发展将为标准制定者带来新的挑战,他们需要将多学科专业知识融合到越来越多的标准项目中。该行业参与这些项目的国际发展也迫在眉睫。BIS、电信工程中心(TEC)和印度电信标准发展协会(TSDSI)之间需密切合作并避免重复工作。

目标 5:明确可以在印度率先开展标准化工作的部门。包括:①在传统优势领域开拓创新标准;②在有时限的项目上制定服务标准;③以研发为基础,开发创新前沿标准。

从历史上看,印度的标准化工作一直遵循其他标准机构制定的标准。印度有许多具有巨大商业潜力的领域,而这些领域仍不在标准化范围之内。在这些领域开展开拓性的标准化工作,不仅会释放其商业潜力,而且会增强印度对全球标准化工作的贡献。这些领域可以包括印度替代药物系统——阿育吠陀(Ayurveda)、瑜伽和自然疗法、乌纳尼(Unani)、悉达(Siddha)、顺势疗法(Homoeopathy),印度美食,印度传统艺术,印度手工艺品,印度传统兽医系统,等等。

为了在诸如航天和生命科学等衍生行业中制定最先进的标准,以推动印度企业成为全球市场上的领先供应商,必须做出有力的努力,在原始研发的基础上制定创新标准。与传统方法不同,这些标准将成为"火炬手"并引领商业发展。

目标6:系统、持续性参与国际和区域标准化工作。包括:①通过自由的资金支持确保候选专家持续参与国际标准化技术委员会工作;②在技术委员会和治理结构中扮演领导角色;③在私营标准的执行中发挥积极作用。

参与制定国际标准的项目能够表达意见和考虑问题,并可能列入国家优先事项和关切事项。随着国际社会越来越多地采用和适应国际标准,这些标准实际上正在成为影响其竞争地位的普遍标准。因此,它们不能包含使印度供应商处于不利地位的条款。这就需要印度专家积极参与国际标准化工作,以了解技术要求及其对贸易、商业的影响。

一个基本的先决条件是在标准项目的生命周期中保持参与的连续性。需要制定一项坚定的政策,确保积极参与每一个确定符合印度利益的标准项目。每个标准化机构、监管机构和相关部门都应确定与之相关的国际论坛,以及对印度参与标准制定或合规机制至关重要的委员会。对于每个论坛,参与计划应与年度会议日程和确定持续参与的合适技术专家相匹配。专家可以来自工业界、科学机构或个体组织。政府应完全承担参与国际标准制定工作的费用,简化程序,并允许根据商定的原则自动供资。

印度应努力在国际标准和项目委员会中担任领导职务,以及在管理机构中获得与印度作为全球主要经济体地位相称的秘书职责。在国际标准、准则和建议书(ISGRs)的发展和审查方面,需注意SPS措施的所有方面,尤其是需要考虑临时采用这些措施的情况越来越多。

随着私营标准的影响越来越大,应按照参与国际论坛的类似方法,参与其他相关论坛。一些SDOs会单独邀请专家参与。这就需要在BIS的支持下制定协调机制,以确保参与这些机构的印度专家对国家优先事项和需求有清晰的认识。南亚区域标准化组织(SARSO)成立于2010年,旨在实现和加强南亚国家在标准化和合格评定领域的协调与合作，旨在为该地区制定统一标准，促进区域内贸易，并加强对南亚区域合作联盟（SAARC）国家供应商的全球市场准入。作为该地区的主要经济体，印度应该代表南亚国家在南亚产品、服务市场准入的标准制定和贸易谈判方面发挥积极作用。除了加强在 SARSO 中的参与度之外，印度还可以在区域合作的标准制定方面发挥主导作用。

目标 7：发展服务业标准。包括：①成立国家工作队，加快服务标准建设工作；②确定服务质量差距、相关基础设施和职业技能所需的标准；③基于差距分析制定国家标准快速程序；④在国际服务标准开发工作中发挥领导作用。

服务业主题在标准化工作中相对较新，在贸易法律文书中也很少涉及。服务的固有特征（如短暂性、异质性、嵌入商品的服务等）增加了该领域标准化的难度。然而，由于服务业在印度和全球经济中都占有很大份额，因此制定服务业标准迫在眉睫。

印度政府已经确定了 12 个优先服务领域（CSS），包括信息技术和信息技术支持服务、医疗价值旅行、运输和物流服务、旅游和酒店服务、会计和金融服务、视听服务、通信服务、法律服务、建筑和相关工程服务、环境服务、金融服务、教育服务，并决定集中资源发展这些领域。

已确定的"优先领域倡议"的干预行动旨在推动这些领域发展的 5 个支柱，其中包括新标准，因为标准在塑造各种服务部门出口竞争力方面的作用正在成为全球服务贸易中日益重要的一个方面。因此，各部门和有关部委的任务是推动制定服务标准，包括所有倡导部门的跨部门横向标准和部门特定的纵向标准，并按照强制性和自愿性通过时限和系统的方式予以采用，以提高竞争力、服务质量和消费者福利。

印度在技能和劳务服务方面的优势有助于其成为许多此类服务的外包目的地。企业家已经在努力加强必要的监管框架、资格标准和机构认证程序，以便根据全球标准对传统服务和新兴服务进行技能开发。该领域至关重要，需要迅速实施，以促进各种服务的出口，并为已确定的优先服务领域创建支持性的技术基础设施，尤其是在主要枢纽周围。

目前在印度，许多服务领域受到法定标准或买方驱动服务协议的约束。在发展较好的领域，现有的服务标准可以符合全球规范，但在大多数情况下，印度服务业的标准水平仍然有待大幅提高。

为了加快标准化工作，应成立一个国家工作队，其任务如下。

（1）在具有相关行业全球经验的专家的帮助下，确定每个一流服务行业的服务质量差距。

（2）制定服务标准，包括所有主要行业的横向标准和特定行业的纵向标准，以解决这一差距，并通过强制性和自愿的途径，有时限和有系统地采用这些标准。

（3）确定服务质量和基础设施所需的标准。

（4）确定制定相关技能标准的职业角色，并创建与之相匹配的包括外语技能在内的培训和人员认证框架。

（5）在信息技术和信息技术支持服务、传统医学系统、瑜伽等领域的标准制定中担任领导角色。

以商务部作为中心机构，形成由 BIS、技能部、优先服务领域的重点部门、领先行业机构、印度质量委员会的代表以及该领域专家组成的工作组。工作组的建议应成为BIS 快速制定服务业标准的重要内容，并由相关部委根据标准提供或升级资源。由于服务业是全球经济尤其是印度经济的主要推动力，因此应早日在服务业标准化的发展中占据领导地位，并努力在国际标准化工作中担任领导职务。

目标 8：创建一个生态系统以迎接来自私营可持续性标准的挑战。包括：①确定影响出口的所有私营可持续性标准方案；②建立包括专家在内的国家应对结构，并在方案标准制定过程中寻求发言权；③在需要时制订相应的计划，开发专家资源，以发展国家生态系统，并促进执行。

在过去几年中，一套被称为私人可持续性标准（PSS）的新标准［有时被称为自愿性可持续性标准（VSS）］已在全球范围内广泛流行。该标准根据可持续发展目标（SDGs）建立在社会进步、经济发展以及环境与气候 3 个基本支柱上。虽然 PSS 对贸易产生了重大影响，但这超出了现有 WTO 制度的范围。它们要么由买方财团推动，要么由致力于可持续发展的机构推动。尽管在许多情况下，特别是对发展中国家的小型运营商来说，成本负担非常高，但供应商别无选择，只能遵守规定。大多数 PSS 都结合了标准、合格评定程序和审核员资格规范。全世界有 500 多项 PSS 在运行，其中有超过 30 项正在影响印度市场。许多 PSS 没有全球利益相关方磋商机制，也不具备参与性或透明性。

因此有必要在国家层面建立一种机制来应对这些标准的挑战。初步任务是确定影响印度出口的所有 PSS，并建立由专家组成的全国响应机制，这些专家可以寻求成员资格或在标准制定过程中发声。

PSS 的一些全球方案是基于符合共同标准和规则的基准方案。只要存在这样的机会,就应该制定和支持至少一个国家计划,以便得到该方案的认可。即使在没有基准的地方,在印度制订相应的计划也能创造必要的生态系统,且便于遵守。此外,还需要在市场上创造资源,以帮助提高行业标准,并提供所需的人手、咨询、培训支持。

3.1.2　合格评定、认可和计量

INSS"合格评定、认可和计量"旨在发展一个可靠、有能力和强健的合格评定基础设施,其使命是为客户和市场提供信心,补充并参与监管监督,从而促进印度出口。

目标 1:提高合格评定方案在国内外市场的可信度。包括:①允许采用多种模式的合格评定程序;②鼓励所有合格评定机构获得国家认可委员会认可;③建立一个市场监督机构来监督法规和自我声明下的一致性声明;④鼓励和承认提供可信的合格评定服务的自愿自我监管机制,作为法规的替代。

印度经济的开放推动了国家质量生态系统的同步发展,目前它包括了一个由第三方检查、产品认证、管理体系认证、人员认证、测试和校准、自我合格声明等组成的完整的合格评定方案。印度的合格评定服务由政府指定的专门组织或私人合格评定机构(CABs)提供,有时甚至直接由监管机构提供。然而,目前仍有相当数量的 CABs 在国家认可委员会的登记之外运营,这些委员会在监督和可计数方面都存在缺陷。由于自我符合性评估水平仍低于标准值,因此完善整个符合性评定的基础设施、激发国内外买家和监管机构的信心显得尤为重要。政府应通过适当的奖励措施和授权做出协调一致的努力,使所有合格评定工作人员都纳入国家认可的范围。

目前,对市场上产品上市后的合规性水平测试手段有限,特别是那些在自我合格声明下的产品。随着越来越多的产品被纳入技术法规,有必要设立一个国家市场监测方案,监测市场上的产品是否符合规定,不论是经核证的还是自我宣布符合规定的。这种监测的结果将为影响评价和合格评定方案的有效性提供有价值的参考。

认证委员会负责确保 CABs 的能力和独立性,其中提高合格评定可信度的补充手段就是建立负责监督合格评定规划和 CABs 完整性和可靠性的自我调节与自我管理机制。通常,负责任的 CABs 和实验室需要开发这样的机制,并对所有运营商施加市场压力,要求它们遵守这些机制。该机制的发展将减少监管并使市场运作更自由。作为替代方案,CABs 在印度的注册和运营可以由产业政策和促进部(DIPP)以适当的方式进行监管。

目标 2：通过商品和服务的认证与部门论坛的相互认可协议，确保并加强全球等效性。

例如，国家认证认可委员会（NABCB）及国家检测和校准实验室认可委员会（NABL）已经在国际认证论坛（IAF）和国际实验室认可合作组织（ILAC）的符合性评定项目认证系列多边认可安排中获得了正式成员资格，从而为进入海外市场创造了更大的机会。这使得通过认证的产品和服务达到市场准入的最低合格标准成为可能。

该协议适用于食品、医药、化工、玩具、医疗器械、电气和信息技术设备、电信设备以及信息技术、教育和技能认证等服务领域。有必要绘制需要满足特定合格评定要求的整个产品和服务范围，并系统建立有助于实现等效状态的设施和基础设施。商务部作为负责贸易的部门，应在该工作中起领导作用，并与各相关部委、监管机构进行协调，通过出口检验委员会、BIS、NABCB、NABL 等有关机构以及出口促进委员会、商品委员会、认证委员会和 CABs 在各自领域推进工作。

目标 3：通过"印度品牌"来促进印度产品被全球所接受。包括：①开发"印度品牌"计划；②印度、国际和进口国家标准认证；③通过出口促进各机构、行业团体、代表团等推广"印度品牌"。

印度产品的推广需要有一个明显可靠的印度品牌认证标签，以确保满足全球范围内认可的质量和可持续实践要求。目前已有的国家认证体系有 BIS 产品认证（ISI）、农产品认证（Agmark）、加工食品认证（FSSAI）、丝绸产品认证（Silk Mark）、家电能效认证等，其中部分体系已存在几十年。然而，就目前的形式而言，它们在地域适用方面受到法律限制，而且仅限于印度标准。为了扩大产品对全球受众的影响，"印度品牌"需要扩大影响力，并且与全球领先的认证相协调。这将需要一个基于国际最佳做法的计划，并对现有国家计划进行基准测试。

这种方法要求进行以下更改：①应按需提供印度、国际和进口国家标准所需的出口产品认证；②产品、工艺和服务认证的多种途径应该符合 ISO CASCO 标准规定。

一旦计划实施，出口促进组织、行业团体、海外代表团应积极推动该计划，并在政府的资助下进行品牌推广。

目标 4：降低合格评定的成本，特别是针对中小微企业（MSMEs），使其具有全球竞争力。包括：①将 MSMEs 的资金援助推广到各种类型的符合性评定；②鼓励海外监管机构和有组织的外国买家接受 NABCB/NABL 与印度国内认可的评定结果等同有效；

③为 MSMEs 建立通用测试设备。

印度迫切需要减轻 MSMEs 在满足监管和海外市场准入要求方面的合规负担。中央政府和州政府一直在提供资金援助,以确保基于 ISO 9001 以及如食品检测、建立实验室的成本质量管理体系认证。由于合规性通常基于测试、检查和产品认证,因此需要在分配的预算条款中将财务援助扩展到所有类型的合格评定。

通过与海外监管机构和买家谈判,接受由 NABCB 和 NABL 提供认证支持的国内测试、检验和认证结果,可以进一步减轻合规负担。能力批准是先决条件,出口检验委员会、出口促进董事会和其他相关组织需要建立健全符合国际模式的合格评定计划,并寻求海外监管机构、计划所有者或有组织的买家的批准,授权他们按照某些部门的类似做法签发合格证书。

鉴于测试设备的成本是中小微企业的主要负担,州政府需要确定共同需求,特别是在集群中存在中小微企业的情况下,协助建立可以在合作或私人基础上运行的通用测试设施。对根据该计划设立的所有实验室都需要提供技术援助,以建立全国可追溯的测量工作并获得 NABL 的认可。

目标 5:积极参与有关合格评定的国际组织。印度是 ISO CASCO 的参与成员。BIS 已经成立了一个镜像委员会,其成员来自认证委员会、行业、SDO、PSUs。CASCO 标准对强制性合格评定计划在全球范围内运行的方式产生了深远的影响。但是,由于文化差异,要求的存在或不存在都会影响这些标准的有效实施。至关重要的是,由相关机构参加 CASCO 的所有工作组并出席其所有会议,以确保提出印度自身的观点并将其纳入已制定的标准中。

由于认证委员会是 IAF/ILAC 的自动成员,因此他们还需要确保参加这些顶尖组织的所有会议。

国家物理实验室(NPL)、国际度量衡局(BIPM)和法律计量学部门的国际法律计量组织(OIML)应确保类似的积极参与。

只要有机会,关键目标应该是担任这些机构的技术委员会、工作组和治理结构的领导职务。

3.1.3　技术法规和 SPS 措施

INSS"技术法规和 SPS 措施"是为了确保印度公民的福祉和安全得到最高程度的

保护。其使命是确保技术法规的目标是实现合法,推进低风险、零负担、高效率的经济发展。

目标1:建立对高效监管实践和监管影响评估的良好理解。包括:①为技术法规以及监管影响评估的发展、实践、审查和修订采用良好的管理实践和政策指导方针;②在负责通知和确保遵守技术法规的机构之间建立协调和理解;③对所有技术法规进行监管影响评估。

由于颁布条例是为了保护和平衡公民社会及各种利益集团的需要,因此这些条例需要根据时代的背景精确地调整风险,承担最小的成本负担,应当易于遵守,并得到透明地管理。它们不应妨碍社会发展和经济增长。技术法规运用不当,可能导致商品和服务价格上涨,产品创新不足,服务质量差。由于规章制度有随时间而失去相关性的趋势,因此需要定期对其进行重新校准,以提高其有效性和目的性。

所有技术法规和SPS措施都应基于良好监管实践的原则,包括基于风险的监管措施选择、对监管效率的考虑,即合规和管理成本与收益之间的平衡,合规的有效性,通报、管理和变更的透明度,沟通的开放性和利益的平衡。

同时,还应评估技术规章对成本效益的影响、对政府的经济负担以及对工业竞争力、市场开放度、小型企业、公共部门和可能受波及的社会群体的影响。建立基于良好监管实践和监管影响评估的政策指导方针,以制定、实施、审查和修订技术规则。

此外,还需要在各部委、监管机构、州政府、执法机构、合格评定机构和社会团体之间建立对遵循良好监管做法和监管影响重要性的全面理解。公务员学院应就这一课题开办认识课程和讲习班。

目标2:分离机构角色以提高效率和避免潜在的利益冲突。由于历史原因,各职能部门、管理机构和合格评定业务的作用已经集中在各自的部门内,其经常出现利益冲突。因此应在体制上逐步分离角色,只保留基本职能,以加强监管框架的有效性。为了实现这一目标,需要将角色分离如下。

(1)部委。①为监管框架和建立相关监管机构(例如FSSAI法案和PNGRB法案)提供必要的授权立法;②为监管框架创建所需的政策和启用规则。

(2)监管机构。①根据适用法案通报具体法规、命令;②建立必要的市场监督(包括港口控制)、合格评定和执法框架,包括进行搜查和扣押操作,以及起诉不合规的制造商和服务提供商。

（3）认证机构。根据监管机构规定的国际标准和要求对合格评定机构进行认证。

（4）合格评定机构。根据监管机构规定的要求,提供独立于监管机构的第三方合格评定服务(认证、检验、测试)。

（5）市场监督机构。①对拟进入或投放市场的产品进行市场前和市场后测试(包括港口控制);②进行搜查和扣押操作,起诉不合规的制造商、服务提供商。

目标 3:确保在全球范围内受到广泛监管的领域得到保护。包括:①确定印度与已通报技术法规的全球惯例之间的差距;②找出印度与技术规范中包含的技术要求(标准/基本要求)的全球惯例之间的差距;③通过系统计划减小差距。

《WTO/SPS 协定》和《WTO/TBT 协定》提供授权考虑,这些事项均来自各个国家或经济联盟发布的技术法规和措施,如人类健康和安全、动植物生命和健康、环境、国家安全,以及通过适当的技术法规或 SPS 措施防止欺骗性贸易行为。基于风险的评估,通报产品和措施。

将印度监管的产品和服务与大多数其他国家进行比较研究后显示,无论是在数量上还是在技术要求上,印度都存在巨大的差距。缺乏技术法规和 SPS 措施会造成在印度生产或者进口至印度的不达标产品对消费者、动植物和环境产生不利影响。通常廉价的不达标产品的流入也影响了经过认证的优良生产商的竞争地位。人们还认识到,最低合规水平可提升部门能力并增强出口潜力。

必须对大多数国家已经发布的,但是印度还没有发布的技术法规和 SPS 措施的领域进行差距分析。在应用良好监管做法和影响评估原则的同时,应将分析作为消除差距的基础;同时在分析 SPS 措施时,还应考虑互惠因素。

还需要对印度技术法规和 SPS 措施中基于国际实践所通报的标准和基本要求进行比较审查,并对其进行适当的修改和升级。

目标 4:技术法规和 SPS 措施的制定应基于适当的标准或基本要求以及与伴随的风险和市场条件相适应的符合性评估程序。包括:①技术法规必须指定最低基本要求,并基于已建立的标准或由专家撰写的技术要求;②必须选择能覆盖风险且不繁重的合格评定路线;③在与各利益相关者磋商以后才能确定。

技术法规和 SPS 措施总是会对产品和服务的生产商和供应商产生合规负担,以及提高整个供应链(包括消费者)和政府执行计划的成本。因此,监管机构需要仔细选择符合法规要求的产品、流程和服务。虽然标准机构专业制定的标准应是首选,但监管机构仍然必须积极主动地参与制定和修订此类标准,并确保其范围符合监管规定的安全保障要求。如果质量标准是针对欺骗性贸易行为制定的,则该标准应包含性能要求,而不应限制输入材料或工艺路线。监管机构还应确保用于监管的标准得到采纳或与国际标准紧密协调。如果基本要求直接纳入技术法规,则应由专业人员提出。

技术法规应选择对行业限制最少和负担最轻的符合性评估程序,以涵盖安全、保安或欺骗风险的广度和程度,并考虑现有的符合性评估机构或实验室基础设施。

目前,依靠合格评定作为合规手段的法规除了少数特殊情况外,还有由 BIS 或监管机构自己运营的选定产品认证路线,这些路线的合规成本很高。由于可以使用其他一些合格评定途径,如自我符合声明、设计认证、批次认证、第三方检验、样品测试、能力审批等,因此在选择之前应综合评估这些途径的最佳适用性。可以优先考虑自我符合声明,但这些应该与上市前和上市后的测试相结合,并以严格的刑事条款来治理故意违约的情况。技术法规只有在利益相关者之间广泛协商并履行 WTO 通报义务后,才能最终确定。

此外,SPS 措施应基于 ISGR。如果 ISGR 不足以达到理想水平,SPS 措施应以科学原理和充分的科学证据为后盾,并以风险评估为基础。如果科学证据不充分,可以根据现有理由暂时采取措施。

目标 5:为技术法规、SPS 措施和符合性评估建立一个全面的监管工具和监督机制。目前,技术法规由不同部门进行通报,其中没有对应部门的法令则由 BIS 负责通报。这些法案是为特定目的起草的,其范围是受限的。根据全球监管惯例和适当的监督及执行规定,为通知标准、技术法规和合格评定程序而制定一项新的授权立法是至关重要的。这项立法应适用于所有那些目前不受特定部门条例管辖的部门,并应规定在必要时设立独立的监管机构。它还应包括关于管制影响评估和定期审查以及日落条款的规定。同时,为了保护消费者免受不合规的供应商或未经认证且不负责任的合格评定机构提出的不合理索赔,适当的监管文书必不可少,同时市场监督机制应具备跨国界信息交换的功能。

目标 6:建立有效的市场监督机制。目前,市场监督活动和其他执法措施由港口的国家政府机构和海关官员处理,然而在技术、资源和授权方面他们并不是最佳人选。由于预计未来对包括网络情报在内的后市场监视和测试的要求将会增加,因此有必要建立一个专业机构来执行或协调市场监视计划(参考符合性评估目标 1)和港口控制操作。市场监督应无一例外地对从市场抽取的产品进行检测，若发现存在故意欺诈行为，则采取法律行动。

目标 7：加强应对海外技术法规和 SPS 措施的机制。新修订的技术法规和 SPS 措施目前由 TBT 和 SPS 成员国定期通报。其中许多对印度的商品和服务的供应有直接或间接的影响。因此，印度迫切需要发展一种动态反应机制，以便在触发预期反应和收到通报后及时做出反应。需要加强机构安排，并建立数字化平台，以分析所有新通报，使其了解或应对监管机构和受影响的供应商（包括潜在供应商），核对他们的需求并及时发布国家对通报的回应。在加强国家应对机制方面，各部委以及行业机构的作用和积极参与至关重要。加强应对系统还将帮助监管机构了解最佳监管实践和监管差距。

发布通告响应包括了解满足合规性要求所需的资源、技术准备和质量基础架构方面的影响。根据影响程度，各相关部门及其下属机构应编制一份影响文件，说明需要政府提供援助，特别是对中小微企业的援助。

3.1.4　意识和教育

INSS 关于"意识和教育"的愿景是在全国范围内打造品质思维，其使命是让每一个公民、组织和机构了解并重视标准化及其相关活动所带来的好处。

目标 1：增强利益相关者的意识。所有公民、组织和机构在生活和工作中都离不开标准、合格评定和技术法规。缺乏对其相关性的认识会导致对标准制定和合格评定过程的参与有限，且无法充分发挥供应商、服务提供者、消费者、决策者和监管者的潜力作用。这一标准化战略的目标是向所有利益相关者广泛传递信息，使他们不仅认识到机会，而且通过积极参与成为负责任的角色参与者，还必须使人们认识到影响国内供应和出口的国家和全球计划实施的具体部门合格性评估要求。

战略中指明增强意识的目标对象应包括负责政策和技术法规的官员，执行机构，中央和州政府的公共采购机构，负责港口管制、贸易和工业的官员和消费者组织。消费者还应该了解在何处以及如何提出投诉、纠纷和上诉。

目标 2：关于标准、符合性评定、技术法规和 SPS 协议的咨询和培训。由于近年来，来自海外进口商的要求日益增多，进口国制定的技术法规和 SPS 措施日益严格，严重影响了印度的出口贸易。这导致了商品和服务的供应商以及进口商在对外贸易中的咨询需求将会越来越大。战略要求 CII、FICCI、ASSOCHAM、FIEO 等贸易组织需要根据这些需求调整其咨询服务,以满足利益相关者的需求。贸易门户网站(由 FIEO 管理)和印度标准门户网站(由 CII 管理)也需要加强其信息中心的作用。

鉴于国际贸易中监管生态系统的复杂性和技术性质,政府官员、监管机构、执法机构和监测机构必须接受充分的培训和技能培养,培训模块应由监管机构、BIS、认证委员会、贸易促进委员会/理事会和顶尖行业协会在州政府行政学院的协助下共同负责。其中 BIS 负责为各层级区域以及产业集群创建培训和意识模块。

专业培训师和顾问在直接向用户传播基于标准的信息方面发挥着重要作用。这项任务最好在非政府部门进行,培训、咨询部门可根据市场需求自由发展。应针对培训、顾问建立一个识别、鉴定系统,并为培训、咨询组织以及不同部门的培训、顾问建立相应的认证系统。印度质量委员会需扩大其管理的标准和合格评定的覆盖范围。

目标 3:为各级教育机构的质量相关内容建立课程。战略认为在潜在劳动力队伍中建立高质量思维的有效任务之一是建立合适的课程,并将其纳入正规教育的各个阶段。这就需要在从小学到高等教育的教育方案中制定与标准化、质量惯例和保护消费者权益的法定条款等有关的具体模块,其中最低限度的教育应该包括影响整个社会的基本标准,如基于 SI 单位的计量学、互操作性的概念,针对人类健康和安全、环境的标准。

随着标准在工程和工业应用中的广泛使用,相关专业课程和标准提供的大量当代技术知识相结合是必不可少的。这将使学生在毕业时"为行业做好准备"。一个标准机构与技术、专业机构之间的紧密联系是至关重要的,为此,需要由人力资源发展部发布政策指南,并将其作为课程认证标准的一部分。

3.2 印度标准化的中长期目标定位

其一,制定和施行国际水准与本土需求相结合的标准及标准化项目。在印度政府不断制定和实施标准的过程中,作为参照或应用的对象,国际标准所占地位日渐提升。相当一部分印度出口产品无法符合国际标准或出口市场的标准,是印度出口发展缓慢的重要原因。近年来,自觉采纳由 ISO 制定的质量管理方面的国际标准,在印度制造商中开始成为一种趋势,他们认识到了在对外贸易中符合国际标准的优点。2015 年,莫迪政府发布 2015—2020 对外贸易政策,其中的核心要点在于提出了未来 5 年印度的出口目标,以及实现这一目标的措施。印度政府寻求至 2020 年,将印度年出口额由当时的 4660 亿美元提升至 9000 亿美元,在世界出口总额中的占比由 2.0% 提升至 3.5%。对于国际标准的采纳成了印度政府致力于实现其极具雄心的出口目标的一个关键配套措施。印度政府鼓励出口产品进行国际标准体系内的质量认证,或直接在国家标准和标准化项目中应用国际标准,以期获得国际市场的认可。与此同时,当本土需求与国际标准存在一定出入时,印度政府仍试图对两者进行调和,如在制定国家标准时,设置了国家标准不必与国际标准相协调的条件,在面向印度国内市场的一些产品认证项目中,则同时采用国家标准和国际标准这两套体系。

其二,参与国际标准制定,将身份由国际标准的遵从者转变为制定者。如前所述,对于向印度出口的外国制造商而言,不少产业领域的印度标准因其与国际标准的差异而成为难以逾越和应对的非关税壁垒,印度因此而长期备受批评。为了同时实现塑造负责任的贸易伙伴形象和保护本国贸易利益的目标,印度近年来开始寻求通过参与多边的标准制定机构,对制定之中的国际标准施加影响,使其更符合印度本土产业的要求和利益。在第四届印度国家标准大会上,时任商务部部长 Nirmala Sitharaman 称,印度应当成为标准制定者,而非仅仅遵从已有的标准,并将如何跻身国际标准制定体系视作印度在标准化问题上面临的两大挑战之一。

其三,建立起一个良性的质量管控和标准运行的生态体系。在提及正在制定之中的国家标准化战略时,印度政府表示,该战略将助力印度建设"一个和谐的、有活力的、成熟的标准生态体系"。在印度的语境下,这一宏观的目标定位具体包括两个主要方面。一是发展出一套有效的管理标准化事务的行政和法律体系。这实际上是印度政府一直以来的目标,印度标准局的建立、《印度标准局法案 1986》及最新的《印度标准局法案 2016》的颁布与生效,都是其重要成果。但对于印度政府而言,这项工作仍然任重而道远。在后文的详细描述中可以看出,印度标准化的事务管理尽管以印度标准局及其法案为一条主线,但在此之外,参与标准化事务的机构和法律规定纷繁复杂,与印度标准局及法案之间构成了一个复杂的体系,但尚未形成严密配合、高效互动的有效机制。这项工作既涉及进一步改进印度标准局的职能及法案,又将涉及协调乃至变动数量众多的其他机构与法案,在印度的行政传统和环境下,这将是一项艰难的工作。二是在中央政府的带动下,在全社会形成主动认可和遵循标准的氛围。目前为止,印度的标准化工作还基本停留在政府主导层面上，而政府的执行力有限，因而始终难以避免市场上大量的标准和认证许可使用混乱乃至违法情况出现。印度政府对此深有体会，已将加强市场管理的内容纳入了《印度标准局法案2016》之中，并在国家标准大会等最新的标准化平台上寻求各利益相关方之间的合作，而良好的社会生态体系的形成并非在朝夕间就可以实现，这将是印度政府在标准化方面的一项长期任务。

第 4 章

印度标准化战略的重点任务及措施

在上述立法体系和管理架构的基础上,印度政府及标准化管理机构从印度在经济和社会方面的关键需求出发,对部分产业领域和项目投入了较多力量,并产出了丰富的标准化成果。本章将先后论述印度标准化战略中的重点产业领域及近年来受到重视的一些新兴项目,展示出印度如何将其经济社会需求与标准化事务结合在一起,并具体以怎样的路径来推进这些产业领域和项目的标准化进程。

4.1　印度标准化战略中的重点产业领域

在印度,标准化事务所覆盖的产业领域非常广泛,仅从印度标准局设立的标准制定部门委员会及其下属的技术委员会的种类即可看出。不过印度政府对于各产业标准化的用力程度仍然有所不同。印度政府最新出台的印度国家标准化战略文件草案指出,目前印度需要优先制定标准的一些关键领域为信息技术、可持续技术与实践、弱势和边缘化部门、工业自动化、机械安全、旅游、医疗保健、教育和技能培训。但在后续的标准化战略发布文本中,删除了印度需要优先制定标准的关键领域。对于印度在信息通信技术领域的标准化情况,本章从标准制定的公私关系的角度分析,后面的章节中将讨论印度信息通信技术领域的新兴重点项目。同时,结合印度目前标准化开展工作情况和标准化战略布局,除信息通信技术领域外,本章还将涉及食品、汽车、医疗保健等领域的标准化情况。

4.1.1　信息通信技术领域的标准化

信息通信技术领域是印度私人部门参与标准制定,以及公私合作的一个典型领域。除以电子及信息技术部及其下属机构为代表的政府部门以外,还存在着多个由私人部门成立或以公私合作(PPP)模式组建的标准制定机构。其中,印度电信标准发展组织(Development Organization of Standards for Telecommunication in India, DOSTI)由印度手机运营商协会等行业组织、高通等在印企业发起;全球信息通信技术印度标准化论坛(Global ICT Standards Forum of India, GISFI)由 CTIF(Centre for Teleinfrastructure)发起,成员中主要包含电信运营商及行业组织,印度电子及信息技术部(Ministry of Electronics and Information Technology)则是支持性力量;印度电信标准发展协会(Telecommunication Standards Development Society for India,TSDSI)则是根据"国家电信政策 2012"的相关设想,按照公私合作模式成立的,政府、产业协会、研发机构、运营商等在其中都有一席之地。

在"国家电信政策 2012"中,印度政府指出,将促进一个包含各利益相关方的电信标准发展组织的建立,以满足安全等方面的国家需求。它将为所有的利益相关方参与国际标准发展组织提供助力,并在推动印度的需求或标准为国际标准所采纳的过程中发挥咨询作用。TSDSI 的成立正是这一愿景的直接结果。在组织架构上,TSDSI 的管理委员会包含了 5 个由印度政府指定的成员,以及 16 个由全体股东选出的成员,前者来自印度电子及信息技术部及其下属的电信部门,而对于后者,TSDSI 将印度电信产业利益相关方划分为 8 个部门,每个部门各有 2 名代表入选,这些部门包括本土制造商、拥有本土知识产权的电信设备制造商、电信服务供应商、政府下属协会社团、学术机构、科研机构、手机或用户端设备制造商、电信软件开发者或服务供应商。在具体的标准制定方面,TSDSI 按照信息通信技术标准制定的具体领域或议题成立了数个研究组(study groups)及协助研究组的工作组(working groups),如目前的研究组包含了无线系统、服务、光纤接入和传输、安全、能源效率,它们的成立决定由管理委员会做出。TSDSI 的任何一个成员都可以向研究组或工作组就某一领域标准化工作的启动进行提议,电子及信息技术部下属的电信部门也可以要求 TSDSI 基于国家利益的考虑而开展标准化工作,这些提议由 TSDSI 管理委员会进行审核并做出决定。上述架构及流程构成了 TSDSI 基本的公私合作模式。在此基础上,TSDSI 代表印度公私部门各利益相关方的利益,参与到国际电信标准的制定之中。例如从 2015 年 1 月起,TSDSI 正式成为 3GPP 的全面组织合作伙伴(full organizational partner),与中国通信标准化协会等其他 6 个国家性或地区性机构一起,共同就 3GPP 的政策和战略等进行决策,扮演国际电信领域标准制定者的角色。

推进标准制定中公私合作的努力不单单出现在信息通信技术等特定领域内。例如在 2016 年,印度质量委员会与联合国可持续性标准论坛(United Nations Forum on Sustainability Standards)共同发起了印度私有可持续性标准国家平台,意在推动核心的政府部门和私营利益相关方之间的定期政策对话,以探讨如何利用和推进私有可持续性标准、应对潜在挑战等议题。印度政府希望加强公私合作,增进印度在制定私有标准方面的实力,并以此应对发达国家当前在制定私有标准、以之控制产业链的主导权,从而维护和增加印度在出口方面的利益。

4.1.2 食品安全领域的标准化

食品安全可谓是印度标准化进程中声势最为浩大、引发最多社会关注的一个产业领域。食品安全直接关涉民生问题,在人口数量巨大、未完全脱离农业社会色彩的印度,食品安全方面的问题极易引发广泛的社会反响,是印度社会重点关注的一个议题。食品安全的标准化由此成为标准化事务的重中之重。在推动的力量上,如前所述,不仅印度标准局下属食品和工业部门委员会涉及这一领域,印度政府还在 2006 年成立了食品安全与标准局这一专门机构,来负责食品标准的制定和产品登记工作,并以《食品安全与标准法案 2006》对食品标准工作进行了详细的法律规定,这样的待遇仅为食品标准领域所拥有,显示出印度政府对于食品标准工作的特别关注。在印度食品安全与标准局的主导之下,印度建立了一个完整而独立的食品安全标准制定、产品检测、认证和登记的体系。

在此基础上,近年来,印度食品安全标准工作进程中有 3 个颇值得关注的方向。一是印度当局寻求加强对于食品市场的监控,规范市场对于食品标准相关规定的遵从。尽管印度政府在食品安全领域投入颇多,但近年来食品安全标准的市场执行状况仍然不甚理想。例如,2012 年,印度食品安全与标准局的调查显示,印度牛奶市场上超过 2/3 的产品不符合标准,有关饮用水安全问题的报道更是层出不穷。该局实际上也已认识到了这一问题,而重点需要解决的是面对印度广阔的食品市场以及薄弱的市场监管机制,印度食品安全与标准局能够投入监管的力量仍显不足的问题。二是印度食品安全与标准局寻求在全国范围内加强食品标准化工作的统筹。尽管目前印度食品安全与标准局是印度食品标准领域最重要的主导部门,但仍然有印度标准局等其他数个部门机构参与食品标准制定领域,相互间的分工配合仍然有疏漏。为此,印度食品安全与标准局为食品安全标准领域的工作制定了一个统一的操作手册,启动了一个信息平台以进行数字检测,以及一个覆盖全国范围的"印度食品实验室网络",以将所有从事食品安全检测的实验室都纳入这个统一的技术平台之中,从而向统筹该领域的工作迈出了重要的一步。三是印度国内的食品安全标准与国际标准的对接成为近年来广受关注的一个问题。近年来,印度食品安全与标准局宣称,印度食品领域的进口量之所以超过出口量,其原因正在于国内食品标准与国际标准的不协调。随着时间的推移,2011 年发布的有关食品安全标准的规定中已有不少内容无法跟上国际食品领域的技术进展、新的食品消费模式等,为此,印度食品安全与标准局在 2013 年启动了使国内食品标准对接国际标准的规划,并在近年来不断重复声明这一愿景。

4.1.3 汽车领域的标准化

印度正在成为世界较大的汽车制造国之一。据统计,在 2015—2016 财年,印度汽车业产值占该年度 GDP 的 7.1%,约占制造业总产值的 49.0%。与此同时,印度的空气污染状况日渐严重,新德里等城市已成为世界上空气污染较严重的城市之一;印度对于能源进口的依赖日趋加重,原油净进口量从 2006—2007 年的 1.115 亿吨增长至 2015—2016 年的 2.0285 亿吨。这些状况都与汽车工业及汽车消费有着密切的关联。与信息通信技术领域的标准化工作相类似,印度汽车领域的标准化也涉及多个政府部门及非政府机构,在汽车尾气排放、汽车燃油效率、汽车安全等多个方面推进印度汽车产业领域的标准化。

汽车尾气排放的标准化由多个部门共同推进,包括中央污染控制局、印度道路运输和公路部(Ministry of Road Transport and Highways)及其下设的排放法执行常务委员会(Standing Committee on the Implementation of Emission Legislation)等。中央污染控制局制定的标准涵盖诸多产业领域,其中涉及汽车领域的主要有以 Bharat Stage 等为代表的汽车排放标准、汽车燃油质量标准、汽车噪声限度标准。其中,自 2000 年启动的 Bharat Stage 计划参照欧盟排放标准为印度汽车尾气排放制定相应标准,并做出时间规划。在该计划中,随着阶段的进展,排放标准趋于严格,从 2010 年开始,印度的 13 个大城市已率先执行 Bharat Stage IV,这一阶段的标准从 2017 年 4 月开始覆盖印度全国,根据规划,Bharat Stage VI 自 2020 年 4 月始在全国实行。道路运输和公路部则根据排放法执行常务委员会制定的标准以及环境与森林部等其他中央政府部门有关排放的规划等,对现有的机动车法案及中央机动车规则中的相关条款做出修订。需要说明的是,Bharat Stage 标准及计划是不断修订中的机动车法案的一部分内容。

电力部(Ministry of Power)及其下设的能源效率局是对汽车燃油效率标准进行规定的主要部门。例如在 2014 年 1 月底,经印度中央政府和能源效率局商定,电力部对外发布机动车的能源消耗标准,根据机动车的生产时间或进口至印度市场的时间,为 2016—2017 财年至 2020—2021 财年,以及 2021—2022 财年及以后这样两个阶段分别设置了制造商需要遵守的平均燃油消耗标准。这一标准的执行则有赖于印度石油与天然气部等部门的配合。

汽车安全标准的制定则主要源于道路运输和公路部下设的汽车工业标准委员会（Automotive Industry Standards Committee）及印度汽车研究协会（Automotive Research Association of India）等。汽车工业标准委员会的成员包含重工业及公有企业部、印度标准局等政府部门，以及中央道路交通协会、印度汽车研究协会等产业界团体；印度汽车研究协会则是由印度各汽车及零部件制造商组成的非政府机构，目前成员数已达 78 个。前者制定汽车安全标准以及印度标准局制定标准中涉及汽车安全的部分，由道路运输和公路部下设的中央机动车规则-技术常务委员会（CMVR-TSC）考虑采纳并对现有的中央机动车规则做出修改。印度汽车研究协会则主要就汽车及零部件的行业制造标准进行具体设定。

4.1.4　医疗保健领域的标准化

与前 3 个聚焦产品的领域不同，该领域主要是服务管理系统的标准化。为提升医疗保健的质量，印度政府在该领域标准化上的投入及形成的标准化成果在印度的服务标准化中处于领先地位。这既是为印度民众提供有保障的医疗保健服务，也意在促进印度医疗旅游的发展，吸引更多外国人赴印进行医疗旅游。该领域的标准化中存在着两个主要的推动力量，并各自形成了一套标准化体系。

卫生与家庭福利部（Ministry of Health and Family Welfare）是其中之一。在 2005 年启动的国家农村卫生任务（National Rural Health Mission）的框架下，印度公共卫生标准（Indian Public Health Standards）于 2007 年由该部发布。而后为了适应新情况，该标准在 2012 年得到更新。在印度公共卫生标准的体系中，印度全国的公共医疗保健机构被划分为 5 个等级，依次是服务一个行政区划的地区医院（district hospitals）、服务 50 万至 60 万人的次区域医院（sub-district/sub-divisional hospitals）、服务 10 万人的社区卫生中心（community health center）、在平原地区服务 3 万人而在部落和丘陵地区服务 2 万人的初级卫生中心（primary health center），以及在平原地区服务 5000 人而在部落和丘陵地区服务 3000 人的卫生分中心（sub center）。针对这 5 个等级的医疗保健机构，印度公共卫生标准分别设定了具体的设立和运作标准，并在全国范围内进行推广。

与印度公共卫生标准相比较，印度质量委员会下设的国家医院与保健机构审定委员会(NABH)设定和推广的标准化体系则主要在两个方面有所差异。其一,国家医院与保健机构审定委员会的标准化工作不止针对公共医疗保健机构,还向私有机构开放。其二,国家医院与保健机构审定委员会的标准化体系以不同的方式对全国的医疗保健机构进行划分,因而制定和推广的标准针对的具体对象也与印度公共卫生标准呈现出显著的不同。国家医院与保健机构审定委员会主要以功能而非服务人口范围来确定其标准的对象,基于此,该委员会设立的标准有"NABH 医院标准""NABH 小型医疗保健机构标准""NABH 血库标准""NABH 血液储存中心标准""NABH 医学影像服务标准""NABH 口腔卫生保健服务者标准""NABH 对抗疗法诊所标准""NABH 健身中心标准""NABH 医院急救中心标准"等。在这些标准的基础上,国家医院与保健机构审定委员会开展了针对各相应类型的医疗保健机构的自愿的资格认定项目。值得重视的是,国家医院与保健机构审定委员会是国际医疗保健质量协会(International Society for Quality in Health Care)的第十二个成员,"NABH 医院标准"已得到该协会的认可,这意味着该标准符合国际医疗保健质量协会所设定的国际医院标准的水平,而根据该标准得到 NABH 认证的印度医院也将得到国际认可。NABH 指出,这一认可有助于推动印度医疗旅游业的发展。

4.2 近年来印度标准化战略中的重要项目

近年来,尤其是在莫迪政府执政以来,为了配合及推进重大的国家经济社会发展计划与方向,印度标准化事务中出现了几个值得关注的新兴项目,它们构成了当前印度特定的国家重大计划和发展方向的一个部分。

4.2.1 智慧城市的标准化

"智慧城市"(smart cities mission)是莫迪政府上台后推出的一个重大项目。自上台伊始,莫迪政府便公布了有关"智慧城市"的计划,在 2015 年 6 月,莫迪政府宣布正式启动该项目。其核心目标在于在全印度促进城市化,正如莫迪所言:"我们不应将城市化视为一个问题,而应将其当作机遇。……我们有责任令城市成为经济增长及消除贫困的中心。"为此,"智慧城市"项目的任务主要包括改造现有城市基础设施和环境、将以信息和数据技术为基础的智慧解决方案应用于城市等。目前为止,共有 90 个城市被纳入建设"智慧城市"的计划,据统计共覆盖了近 9600 万城市人口,预期将花费约 18925 亿卢比(约合 259 亿美元)。

为了推进"智慧城市"项目,"智慧城市"建设的标准化几乎在该项目启动的同时便被提上了议事日程。就在该项目启动的当月,就有消息称,印度标准局将研究如何为"智慧城市"的建设提供标准化的指导方针。随后,印度标准局在标准咨询委员会的土木工程部门技术委员会之下成立了智慧城市委员会,以专门负责"智慧城市"标准的起草工作。该技术委员会与印度住房和城市事务部密切合作。2016 年 9 月底,印度标准局发布了"智慧城市"衡量指标的草案,并为该草案征求公众意见。

根据上述草案来看,这是一个综合性的标准体系,根据"智慧城市"计划的总体需求来确定其所覆盖的诸多具体领域。在草案中,所有的标准按照 17 个具体领域进行划分,包括经济、教育、能源、环境、财政、防火及应急响应、政府管理、健康、娱乐、安全、避难所、固体废物、电信及创新、交通、城市规划、下水道系统及公共卫生、水源供应。可以看出,这一标准化体系不仅限于一般的产品标准范围,而且极力向服务、管理等领域拓展,几乎涵盖了城市化所涉及的一切议题。通过设置指标群的方式,该标准化项目意在使"智慧城市"的建设方向量化。对于上述每一个领域,印度标准局都为之设定了一定的核心指标和辅助性指标。综合全部领域而言,共有 46 个核心指标和 47 个辅助性指标。

总体而言,印度"智慧城市"的标准制定既在很大程度上参考和仿照了 ISO 的相关标准体系,也对印度本土的需求进行了考量和采纳。这显示了印度"智慧城市"项目是将现代化与印度特色相结合的尝试。在指标的领域和范围划分上,即上述 17 个具体领域,印度的标准大体上沿用了 ISO 发布的"社区可持续发展——城市服务和生活质量的指标"这一国际标准系统的模式。就具体领域中的指标而言,则与 ISO 的指标有同有异。例如,在印度为经济领域所设的 4 项核心指标中,失业率、工商业的估定价值在总估定价值中的比例与 ISO 标准相同;与此同时,印度放弃了"城市人口的贫困率"这一项指标,而另以"本地生产总值""人均生产总值"这两项作为核心指标,显示了作为新兴经济体的印度对于经济增长的关注。又如,关于安全领域的核心指标,除了 ISO 标准所列出的"每 10 万人的警察数量""每 10 万人的杀人犯数量"这两项之外,印度的标准中还包括了"每 10 万人中针对女性的犯罪数量",显示了印度政府解决其特殊的安全问题的意图。再如,在电信与创新领域,ISO 设立了两项核心指标,即"每 10 万人的网络接入数量""每 10 万人的移动电话接入数量",而在印度的核心指标中,除第一项相同之外,另两项则为"拥有电脑的家庭数量的比例",以及"信息通信基础设施及在线市民服务的网络安全就绪情况",这反映了当前印度政府积极普及网络接入、提升网络服务、增强网络安全的尝试。

4.2.2　网络安全的标准化

近年来,网络安全是印度政府关注的一个焦点领域。2013 年,印度政府发布了《国家网络安全政策》(*National Cyber Security Policy*),旨在将维护网络安全提高到国家战略的高度上,构建一个维护网络安全的政策体系。网络安全的标准化在其中发挥着重要作用。在该政策文件中,印度国家网络安全的政策目标的第二项即围绕标准问题展开:建立一个保障体系,推进安全政策的设计,以及通过产品、流程、技术及人员的合格评定促使网络安全行动符合国际安全标准和最佳实践;在具体的战略中再度强调,国家的网络安全政策需为安全的信息流通、危机管理计划、主动的安全评估及信息基础设施建立标准和机制。

目前来看,印度政府正在至少两个层面上积极推进网络安全的标准化。其一,印度政府正在制定专门的网络安全技术标准。印度电子及信息技术部是积极推动这一进程的主要力量。据报道,2017 年该部正在制定移动电话的网络安全标准,并即将完成。在此过程中,该部已向诸多移动电话的制造商发布通知,要求这些制造商提供相关信息,以说明他们采取了何种安全措施以保障所生产的移动电话的网络安全,这里移动电话的网络安全主要包括防止用户数据隐私的泄露、防止移动电话的数据被监控等。其中,相当一部分是来自中国的手机制造商。实际上,印度政府最为担忧的是本国的数据信息为他国所获取,而由他国制造商生产的移动电话被视为一个极大的安全隐患,相关网络安全标准的出台则是为了在技术上对这些设备的安全性进行把控,它构成了印度在投资上审查外来电信运营商、要求数据存储本地化等一系列网络安全措施中的一个重要环节。

其二,印度政府正在制定和颁布相关法律,以强制执行网络安全标准方面的产品和服务认证。在印度电子及信息技术部看来,既有的相关法案,无论是"印度标准局法案"还是"印度电报法案",都未能为网络安全方面的强制认证提供法律基础,这是既有法律体系的一个缺陷。为此,需要建立新的专门法案,赋予网络安全的认证以强制性的法律地位,从而以法律的手段强制相关方必须遵守印度政府设立的网络安全标准。

4.2.3　印度标准化战略的"绿色计划"

随着工业化的发展,作为新兴经济体的印度面临的环境挑战日益严峻。无论是印度城市空气质量的恶化还是能源的可持续,都是印度亟待解决的重要问题,而不能只顾及经济增长,无视其环境后果。前述莫迪呼吁的印度制造的"零缺陷、零副作用"中,后一层含义就是希望印度制造环境友好型产品。为此,近年来,在国际舞台上,印度积极参与到太阳能等可再生能源的倡议中,2016 年巴黎峰会期间与法国共同发起了国际太阳能联盟。在印度国内,标准化也成了推进开发和利用可再生能源的重要方式。

2014 年以来,印度标准局明确提出要启动印度标准化的"绿色计划"(Green Initiative)。根据规划,该计划应当包含多个具体的项目,目前已向公众公布并开始实施的项目是屋顶太阳能电站项目(Rooftop Solar Power Plants)。作为示范,印度标准局率先行动,在其总部大楼及各地区办公室的大楼楼顶安装屋顶太阳能电站。据印度标准局网站显示,目前为止,已完成了 394.4kW 太阳能电站的安装,如位于新德里、金奈的办公室大楼都分别安装了 100kW 的太阳能电站,而且公众可以通过网络对这些已安装的太阳能电站的运行情况进行实时监测。不仅如此,印度标准局还在寻求将这一项目向印度国内的其他办公楼等拓展,以助力印度政府提出的太阳能安装目标的实现。

此外,印度有关太阳能产业的具体标准及认证的相关规定近年来也得到了不断地更新。在产业标准方面,印度新能源及可再生能源部(Ministry of New and Renewable Energy)于 2014 年更新了有关太阳能产业的标准和技术规定,将太阳能产业中更多产品种类及其相应标准纳入了其中。实际上,该部在太阳能产业领域并不自己制定标准,其要求印度的太阳能制造商在各种具体产品上所遵循的标准大部分来自 IEC 标准,也包含了少部分的"印度标准"。不过,这实际上是对于相关制造商自愿符合标准的要求,而非强制认证计划。2017 年 8 月底,印度新能源及可再生能源部发布行政令,名为"太阳能光伏、系统、设备及组件(强制认证要求)行政令 2017"[Solar Photovoltaics, Systems, Devices and Components Goods (Requirements for Compulsory Registration) Order,

2017]，为将太阳能产业相关产品纳入强制认证范畴提供了正式的法律依据。该行政令于发布的一年后正式生效。

4.3 印度标准化战略的对内重要措施

为了推进标准化进程，除标准制定和认证项目等基本的工作以外，印度政府还致力于采取诸多措施，为其标准化的实施提供支持和保障。在对内方面，印度政府旨在纠正国内产业界、消费者等各相关方对于标准化缺乏认知的状况，提升社会经济活动中追求产品和服务质量的氛围，并使标准化的执行和实施更为便利。不仅如此，印度政府还关注对外合作，在多边和双边的层面上参与国际标准化合作，提升印度在国际舞台上的影响力和话语权，加强其他国家对于印度标准化的认可，并切实以标准化互动带动对外贸易。

4.3.1 开展对于产业界的培训

标准化战略的执行和实施离不开产业界对于标准化及其相关流程的了解与应用。由此，包括印度标准局在内的相关机构组织纷纷开展对于产业界的标准化培训。具体而言，这些培训主要面向两类人员：一是生产商，目的在于令其更为准确、熟练地将标准化管理应用于生产活动；二是行业内的标准发展组织与相关技术人员，培训的目的是提升其发展行业标准的能力。

对于标准化培训，印度标准局已建立起了一个较为完整的系统。印度标准局下设的国家标准化研究院（National Institute of Training for Standardization），专门负责标准化培训事务。该研究院在新德里诺伊达地区建有主校园，并在金奈、加尔各答、孟买、班加罗尔等地设有培训中心。国家标准化研究院开展的标准化培训项目包含了国内和国际两大块。在国内培训方面，国家标准化研究院每年召开多场内部的或公开的培训项目，涵盖了管理系统、标准化促进、金银首饰认证、实验室管理、印度标准局许可获取流程等方面。以针对行业内标准发展组织与相关技术人员的标准化培训系列项目为例，为标准发展组织召开的培训项目意在使这些组织理解标准化的相关理念、统一行业标准制定的流程、提升他们参与国家或国际标准化进程的意识；面向技术委员会成员的培训项目则意在提升相关人员对于标准制定的运作流程及其意义的了解。又如有关印度标准局产品认证的培训项目,既面向需要获取认证的生产商开展,使其了解获取许可和认证的流程、费用情况、使用印度标准局网站的方法、误用印度标准局标志的后果等知识,也面向已经获取了许可的生产商,使其了解使用许可的要求、应对检测和检查的需求等。单 2015 财年,印度标准局就开展了 111 项国内培训项目,而对于下一年度的培训项目,印度标准局既会提前就公开培训项目做出规划,也会接受生产商等团体的要求,开展特定的内部培训项目。2021 财年,印度确定开展的线上线下培训多达 116 项。与此同时,印度标准局还有着面向发展中国家开展国际标准化培训的传统,具体培训项目包括管理系统、标准化与质量保证、实验室质量管理系统三类。例如在标准化与质量保证培训项目中,具体培训内容涉及标准化流程、认证、检验、数据技术等方面的教学以及实地参访等。在技术援助计划之下,印度政府还根据一定条件为部分参加培训的国外人员提供助学金,以鼓励人们前往印度参加此类培训。

除印度标准局外,还有其他非政府机构积极参与到标准化的培训工作之中。如印度工业联合会(Confederation of Indian Industry)下设的质量协会(Institute of Quality)即是其中的一个重要力量。质量协会旨在通过提升印度工业界在提供标准化产品和解决方案方面的认知与能力,促进印度工业的竞争力,手段主要包括开展培训与研讨会、最佳实践分享、评估等。在培训方面,质量协会开展的质量管理系统培训与标准化事务直接相关,其主要内容有帮助相关组织机构建立质量战略框架、基于风险模式建立商业质量体系、在复杂环境中应用有效的流程管理框架,为相关组织机构提供质量应用方面的差距分析和评估服务,帮助合格的评定机构获取基于 ISO 标准的国际认可,为中高层质量管理者提供有关质量管理的证书课程,等等。

4.3.2　提升公众对于"印度标准"的认知

在印度,公众对于"印度标准"的认知仍然有限。根据相关调查研究,绝大多数的民众缺乏对于 ISI 标志的认知,因而更是无法在日常消费中对带有 ISI 标志的产品做出有意选择。因此,提升公众关于"印度标准"及其产品认证的认知对于印度政府的标准化管理而言是一项重要的工作。

印度标准局的日常工作内容之一就是策划和运作有关加强消费者意识的活动和项目。其活动主要是通过丰富多样的媒介,如纸媒、电子传媒、户外活动等对"印度标准"

及 ISI 标志进行宣传。如在 2015 财年,印度标准局举办了 166 次宣传活动,具体形式包括在电视节目、广播和报纸杂志中广而告之,在地铁、机场、火车站等公共场所的电子牌上进行展示,在银行存折上印刷相关内容,在电影院穿插"印度标准"的介绍,等等。此外,印度标准局还启动了标准使用教育项目(Educational Utilization of Standard Programs),以期在青年学子之中宣传"印度标准",使之更好地接受有关标准化和质量的理念。不过,印度标准局也认识到,仅凭这些活动,很难全面地接触到印度的广大人口,使有关标准化的认知真正地普及开来。为此,印度标准局启动了消费者组织认可计划(Scheme of Recognition of Consumer Organizations),意在与消费者组织建立联系,将其作为向民众进行宣传的中介,从而加强政府与消费者之间的联系。该计划设定了相关消费者组织加入的标准,受到印度标准局认可的消费者组织在向消费者宣传印度标准局的活动、购买 ISI 标志的产品的益处、强制认证的产品范围、误用 ISI 标志的后果及投诉渠道等方面负有一定责任。同时,不仅是印度标准局及消费者组织,政府机构如印度质量委员会下属的国家质量促进署(National Board for Quality Promotion)等也致力于在消费者中传播有关标准和产品认证的知识。

这一类活动和项目的主要目的有二:其一,对普通消费者的消费行为进行引导,从而形成有关标准和质量的良好市场氛围;其二,帮助和鼓励消费者就 ISI 标志误用等情况进行投诉,从而加强对制造商的标准化管理。

4.3.3 推进印度标准化的电子管理

近年来,在"数字印度"(Digital India)计划的风潮之下,对于标准化相关事务的电子管理已被提上了议事日程。已经启动的行动包括开通相关网站及标准化领域的网上办事通道等,既可以节省政府标准化管理的资源和成本,也可以使各相关方之间的信息沟通及相关行政流程更为便利、更具效率。

目前为止,印度政府进行标准化的电子管理的主要措施有二:其一,开通在线认证服务系统,使产品认证的绝大部分流程实现数字化。在印度标准局的主导之下,有关在线认证服务的专门网站(MANAK Online)得以开通(manakonline.in)。通过该网站,在注册账号之后,有进行产品认证需求的生产商可以在线申请产品认证。在收到申请后,该系统可以为申请者自动分配产品认证的检测官员以及检测日期,申请者则可以在线对审核进度进行跟踪,并获取相关报告。这一在线系统覆盖了印度标准局所设立的所有的标准认证计划,除了许可的颁发这一行为无法通过在线系统完成以外,通过在线认证服务可以实现许可的更新、撤销、暂停使用及监督等多项流程。此外,该网站还公布了已颁发的认证许可,并允许任何个人用户根据一定的规定和资质申请成为相关认证计划的审核员。可以看出,通过该项目,印度标准局的官员等相关行政人员及认证许可的申请者都可便利地参与到标准认证之中,整个标准认证流程的运作效率、标准认证的公开性和透明度都得到了极大的提升。

其二,开通和建设"印度标准门户"(India Standards Portal),从而建立起一个统一的有关印度标准化体系的信息平台。该网站(indiastandsportal. org)是在 2017 年第四届国家标准大会上正式启动的,目前仍处于持续的内容建设之中,是一个包含了标准与技术规定、合格评定及资格认证实践等所有相关信息的一站式平台。在基本内容方面,该网站列出了印度国内参与标准化工作的重要机构,包括标准制定机构、审定机构、得到许可的认证机构及检测实验室、其他支持性组织等,并对它们一一做出详细介绍。此外,在该门户网站上还可查到来自印度及其他国家的 TBT/SPS 通报。

4.3.4 召开国家标准大会

自 2014 年开始组织召开印度国家标准大会。该活动由印度商业和工业部(Ministry of Commerce and Industry)、印度工业联合会以及印度标准局等多个机构部门共同主办,参与者则涵盖了印度标准化的各利益相关方,包括中央及地方邦政府、标准制定与合格评定机构、企业等。

国家标准大会召开的核心目的至少有两个方面。一者,提升各方对于标准化的重视程度,从而为印度标准化的推进凝聚更多力量。在第二届国家标准大会上,时任印度商工部商务秘书的 Rajeev Kher 指出,印度的标准化工作面临着数重挑战,其中首要的便是如何令各利益相关方切实认知到标准的重要性。印度政府在其就第四届大会发布的新闻稿中表明,大会旨在令产业界、各级政府部门及标准制定与合格评定机构认识到变化之中的国际贸易背景下标准日益增长的重要性并为此做好准备。为了提升这种认知,国家标准大会关注其他国家的标准化经验,并分产业领域讨论印度在标准化进程中面临的机遇与挑战。二者,为国家标准战略的制定等工作创造一个国家性的、集合各方力量的平台。在 2017 年的国家标准大会上,国家标准战略的制定被提上了议事日程,成为会议的一个重要主题。本届会议所设置的分会议中,第一个即是"国家标准战略的主题文件",聚焦正在起草的这一战略进行讨论。未来将要发布的这一战略意在助力印度发展一个和谐的、有活力的、成熟的标准生态体系。

4.4 印度标准化战略的对外重要措施

4.4.1 印度在 ISO、IEC 中的参与

在 ISO 和 IEC 中,印度的表现日渐活跃。近年来,在 ISO 和 IEC 的一些技术委员会或分委员会中,印度日渐从观察员国身份转换为参与国身份。从印度最近几年在这两个国际组织的技术委员会和分委员会中分别作为成员和观察员的数量变化趋势可以看出,无论是在 ISO 还是 IEC 中,印度以参与国身份参与技术委员会或分委员会日渐增多,在两种身份的总数无固定增长趋势的情况下,观察员国身份向参与国的转换不言而喻(见表 4-1)。这种转换反映了印度在 ISO 和 IEC 中发挥更大作用、把握更大话语权的意图。

表 4-1　印度参与 ISO/IEC 技术委员会或分委员会的数量

截止时间	ISO			IEC		
	P 成员	O 成员	总数	P 成员	O 成员	总数
2020.07.31	473	194	667	98	72	170
2017.10.31	436	218	654	87	73	160
2016.03.31	418	244	662	80	76	156
2015.03.31	416	247	663	74	79	153
2014.03.31	348	278	626	70	85	155

资料来源:2017 年、2020 年数据根据 ISO 及 IEC 官网整理;2014—2016 年数据根据印度标准局 2013—2014 年以来的年报整理而得。在此前的年报中,相关数据缺失。

不仅如此,印度更是通过参与 ISO 与 IEC 中关键的政策发展委员会、管理委员会,以及组织相关技术委员会的会议等方式,在其中提升自身的影响力。例如,印度标准局在 2013—2015 年曾参与 ISO 的技术管理委员会(Technical Management Board),当前则是 IEC 标准管理委员会(Standards Management Board)2015—2017 年任期的成员。

印度还在 ISO 的一些技术委员会中承担秘书处之职,截至 2020 年 7 月,印度标准局正在担当秘书处的 ISO 和 IEC 技术委员会有 11 个,分别是:ISO/IEC JTC 1/SC 7 软件与系统工程、ISO/TC34/SC7 香料、烹饪香草和调味品、ISO/TC 113 明渠水流测量、ISO/TC 113/SC 1 流速面积方法、ISO/TC 113/SC 6 沉积物运送、ISO/TC 120 皮革、ISO/TC 120/SC 1 原料皮(包括含浸酸皮、ISO/TC 120/SC 2 鞣制革、ISO/TC 120/SC 3 皮革产品)、ISO/TC 146/SC1 固定源排放、ISO/TC 332 金融机构和商业组织的安全设备。

2006 年,印度与国际标准化组织签订了"印度标准局(BIS)—国家标准化研究院(NITS)和国际标准化组织(ISO)"的谅解备忘录,涉及的主要内容如下:认识到国际标准化组织是处理标准化事项的专门国际机构,其目标是促进国际货物和服务交流,发展知识、科学、技术和经济活动领域的合作,包括通过推广国际标准向发展中国家转让技术和良好的商业做法。印度标准局是印度消费者事务、食品和公共分配部(Ministry of Consumer Affairs, Food and Public Distribution)下属的一个法定组织,从事以下活动:①制定"印度标准";②认证(产品和管理系统);③产品测试;④世贸组织咨询点;⑤培训。

根据双边讨论的结果,国际标准化组织秘书长和国际标准化组织总干事商定了以下谅解备忘录,以指导两个组织之间的合作,并协调它们在其职能和活动相辅相成的所有领域的活动。备忘录包含谅解备忘录的作用、合作领域、推广和信息交流、行政和财务模式、最终条款 5 个部分的内容。

4.4.2　印度在区域多边标准化合作体系中的参与

在区域多边标准化合作的层面上,印度是太平洋地区标准大会(Pacific Area Standards Congress)及南亚区域合作联盟(SAARC)框架下的南亚区域标准组织(South Asian Regional Standards Organization)的重要参与方。此外,印度巴西南非对话论坛(IBSA)这一三边合作平台在推进标准化合作方面也有所成就。

具体而言,太平洋地区标准大会与南亚区域标准组织都是专门的区域性标准化组织。前者于 1973 年正式成立,现有 24 个成员,覆盖了亚太地区的大部分国家。太平洋地区标准大会每年召开一次全体会议,2015 年,第 38 次大会即由印度主办召开。作为南亚区域合作联盟之下的一个专门机构,南亚区域标准组织于 2011 年正式成立。该组织不仅为各成员国商讨在标准化问题上的共识提供了平台,而且就 5 个具体的产业领域设立了部门技术委员会,以制定和发布 SAARC 地区标准,这些领域包括食品与农产品,黄麻、纺织及皮革,建筑材料,电子及信息通信,化学及化学品。

IBSA 则于 2003 年成立,是覆盖多领域议题的跨区域多边合作平台,标准化正是其中的一个合作议题。在该议题上,IBSA 的合作成就包括 2006 年成立标准、技术规则及合规评定小组,专门负责三国间的标准化合作事项,并于同年发布关于标准、技术规则及合规评定的贸易便利化行动计划,该计划确定了三国间标准化合作的几项基本目标和途径,如建立各国间信息及经验交流机制,组织聚焦具体产业领域的研讨会,探讨在国际标准制定中协调立场、互相帮助的可能性,建立有关标准与技术规则互认的信任建立措施,探寻各国间标准对接的可能性;2008 年,三国签订关于标准、贸易便利化、技术规则及认可的谅解备忘录,进一步确认在标准化方面进行信息交流、建立信任措施及协调标准等合作的具体措施。此外,三国的中央标准化机构还定期召开会议,商讨合作。

通过参与这些区域性的多边标准化合作,印度意在促进对外贸易,更好地融入区域产业价值链。2015 年,由印度主办的太平洋地区标准大会第 38 次会议的主题为"服务业的标准化战略:当下的工作与未来的倡议"。服务业在印度的经济增长及出口中都占据着主导性的地位,如在 2016—2017 财年,服务业总增加值占据三大部门总和的 53.66%,该年度服务业出口则约占印度总出口的 58%。在上述大会上,印度消费者事务、食品和公共分配部的秘书指出,随着服务业贸易的不断增长,教育、信息通信技术、旅游、零售、物流等领域的标准化变得非常重要。印度近年来一直希望拓展在亚太地区的服务业出口市场,以满足本国服务业的出口需求。而南亚区域标准组织制定共同标准的行动,更是被称为迈向创造南亚地区产业价值链的第一步。在南亚区域标准组织成立以前,有印度学者指出,南亚地区标准化的缺失阻碍了各国间相互的产品出口,促进南亚区域经济一体化亟须在各国间协调标准。IBSA 达成的有关标准化合作的谅解备忘录则直接指出,三国认可标准领域合作对于防止和消除贸易的技术壁垒、促进相互间贸易流通发挥着非常重要的作用。标准化合作正是 IBSA 在贸易领域所达成的少数几项切实成果之一,显示了印度等国在考虑贸易促进手段时对于标准化问题的重视。

4.4.3 印度与其他国家的双边标准化合作

以印度标准局为主体，印度政府与不少国家对应的标准化机构签订了合作协议，包括谅解备忘录（MOU）与互认协议（MRA）。印度标准局已达成的双边合作如表 4-2 所示。截至 2020 年底，印度标准局共与 30 个国家的 33 个机构签订了 MOU 或 MRA。下面将简要概述印度标准局与世界重要经济体签订的标准化合作谅解备忘录。

表 4-2　印度标准局签订的双边 MOU 与 MRA

地区	国家	序号	合作机构	性质
亚洲	阿富汗	1	Afghan National Standardization Authority	MOU
	孟加拉国	2	Bangladesh Standard and Testing Institution	MOU
	不丹	3	Royal Government of Bhutan	MOU
	巴基斯坦	4	Pakistan Standards & Quality Control Authority	MRA
	斯里兰卡	5	Sri Lanka Standards Institution	MRA
	吉尔吉斯斯坦	6	Ministry of Economy of Kyrgyzstan	MOU
	乌兹别克斯坦	7	Agency for Standardization, Metrology and Certification of Uzbekistan	MOU
	伊朗	8	Institute of Standards and Industrial Research	MOU
	以色列	9	Standards Institution of Israel	MOU
	阿曼	10	Directorate General for Standards and Metrology	MOU
	阿联酋	11	Emirates Authority for Standardization and Metrology	MOU
	日本	12	Japanese Industrial Standards Committee	MOU
	越南	13	Directorate for Standards, Metrology and Quality	MOU
	约旦	14	Jordan Standards and Metrology Organization	MOU
大洋洲	斐济	15	Department of National Trade Measurement and Standards	MOU
欧洲	法国	16	Union Technique de l'Electricite	MOU
	德国	17	DIN (Deutsches Institut fur 8. Normung)	MOU
		18	DKE German Commission for Electrical, Electronic and Information Technologies of DIN & VDE	MOU
	希腊	19	Hellenic Organization for Standardization	MOU
	斯洛伐克	20	Slovak Office of Standards, Metrology and Testing	MOU

续　表

地区	国家	序号	合作机构	性质
欧洲	乌克兰	21	The State Committee of Ukraine for Technical Regulation and Consumer Policy	MOU
	斯洛文尼亚	22	Slovenian Institute for Standardization	MOU
	俄罗斯	23	Federal Agency on Technical Regulation and Metrology	MOU
	白俄罗斯	24	State Committee for Standardization of the Republic of Belarus	MOU
美洲	美国	25	American National Standards Institute	MOU
		26	American National Standards Institute,与 ANSI 及印度工业联合会(CII)的三方协议	MOU
		27	Institute of Electrical and Electronics Engineers	MOU
	苏里南	28	Institute of Suriname Standards Bureau	MOU
非洲	埃及	29	Egyptian Organization for Standardization	MOU
	加纳	30	Ghana Standards Bureau	MOU
	毛里求斯	31	Mauritius Standards Bureau	MOU
	尼日利亚	32	Standards Organization of Nigeria	MOU
	马里	33	Direction Nationale de Industries	MOU

注:序号只是为了标示合作的数量,并不表示达成合作的时间顺序。

1. 印度标准局与美国相关机构签订标准化合作备忘录

通过这些双边协议,印度标准局切实地推进了与其他一些国家的标准化合作。以美国与印度的合作为例,2006 年,印度标准局与美国国家标准协会(ANSI)签订 MOU,通过该文件,双方同意就国际及区域标准化组织的活动、参与标准化的审定机构体系、双方已发布的国家标准及流程名录、合格评定、标准化培训等诸多内容交换意见和信息,并共同促进两国标准化领域专家、标准机构等之间的交流。关于标准化、合格性评估以及各组织之间相互交流信息和出版物等事项达成的条款如下。

(1)印度标准局和美国国家标准协会(ANSI)同意:①就各种国际和区域标准化和合格评估组织的活动、业务和工作方案交换意见,包括但不限于国际标准化组织和国际电工委员会;②交流彼此的观点,并在可行的情况下,努力就这些组织对其各自成员的作用制定相互支持的立场;③就参与标准化和合格评估的机构的认证制度交换信息。

(2)印度标准局和美国国家标准协会还同意:①促进印度和美国专家在标准化、合格评估和质量保证等各个领域的对话，这是由双方之间的特殊安排决定的；②鼓励本国的标准组织确定其对应群体，以建立双边关系，包括相互交流信息和材料、贸易展览等；③安排印度标准局和美国国家标准协会高级管理层的互访等。

（3）印度标准局和美国国家标准协会同意使用所有适当的媒体进行交流：①各自组织按照其国家标准和程序发布的当前目录；②各自组织发布的有关合格评定的一般信息和出版物；③标准化培训课程的一般材料，各自组织制定和发布的合格评定和质量保证；④各自组织发布的定期出版物和其他相关信息。除此项目外，印度标准局还被授予美国国家标准协会的翻译和重印 ANSI Reporter 任何版本的免费文章的许可。

（4）由本谅解备忘录的解释或执行引起的任何问题，将通过印度标准局和美国国家标准协会之间的协商或他们可能共同决定的其他方式解决。

（5）本谅解备忘录可通过书面协议或双方交换照会予以修正；任一方均可通过至少提前 6 个月向另一方发出书面通知来终止本谅解备忘录；本谅解备忘录在其有效期届满之时，对正在执行的项目仍然有效。

一年之后，认识到在标准和符合性评估方面开展全球合作的可取性，为促进在印度和美国分享有关标准、符合性评估、临床规范（SCATR）和其他与贸易有关的信息，同时加强对每个市场准入要求的了解，印度标准局与美国国家标准协会加上印度工业联合会（CII）签订了一份三方的建立印度—美国标准门户网站的谅解备忘录，主要内容包括通过建设印度—美国标准门户网站的方式进一步促进两国间有关标准、合格评定等信息的交换，为两国的企业等利益相关方提供获取相关信息的平台。达成的谅解条款如下。

（1）印度标准局、印度工业联合会和美国国家标准协会同意建立公共标准门户网站，共享与以下内容有关的信息：标准、合格评定、技术法规、贸易。

（2）该门户网站将包括一个单一的开放网页，提供详细的门户目标，并简要概述印度标准局、印度工业联合会和美国国家标准协会。该门户网站将包括与三者网站以及印度和美国其他相关组织的超链接，包括标准制定组织、从事合格评定的机构、技术条例、政府组织、行业协会和贸易促进组织。美国国家标准协会将确定包含门户的美国部分的适当链接，印度标准局、印度工业联合会将确定包含门户的印度部分的适当链接。

（3）印度标准局、印度工业联合会和美国国家标准协会同意：①在美国和印度，以及在区域和国际组织中，交换对各种工作计划的活动中期运作和形式评估的看法；②支持和鼓励双方经济技术专家的合作参与;③确定和促进印度标准局、印度工业联合会和美国国家标准协会各自成员和选区内的对应群体之间的沟通;④共同努力澄清并获得彼此经济的市场准入要求;⑤在适当的情况下,在所有上述情况下,为彼此的成员寻求和分享商业机会。

(4)本谅解备忘录的解释和执行所产生的任何问题将由印度标准局、印度工业联合会和美国国家标准协会之间的相互协商或共同决定等方式解决。

在此基础上,2009年,美国—印度标准及一致性合作项目(U. S.-India Standards and Conformance Cooperation Program)启动,该项目具体包含了3项主要内容:①分别创建并发布美国标准名录和印度标准名录,提供有关美、印两国的市场准入的标准信息;②在印度召开一系列针对具体标准领域的工作坊,召集两国的产业界和技术专家以探讨标准规定问题;③在ANSI标准门户网站(ANSI Standards Portal,www.standardsportal.org)上提供有关美、印两国标准及认证体系方面的相关信息。最后一点实际上就是2007年三方MOU中提出要建设的印美标准门户网站。ANSI标准门户网站的下设页面中提供的信息有面向美国对印出口商的印度标准名录,印度标准、认证、税务等方面的信息介绍,面向印度的美国商业指南,等等。不过目前看来,这一信息平台上的信息主要是为美国出口商服务的,事实上在整个美印标准化合作的过程中,美国都是更具积极性和主动性的一方,显示出美国希望打开印度市场、获取更多经济利益的强烈愿望。

2015年,印度与美国电气与电子工程师协会(IEEE)之间关于标准化合作达成谅解备忘录。印度标准局于1987年4月1日通过1986年11月26日的议会法案。印度标准局是一个由印度消费者事务、食品和公共分配部管理的机构,IEEE是世界上最大的工程学会,其成员遍布160多个国家,专注于推进电气、电子和计算机工程、计算机科学和相关技术的理论和实践。IEEE标准协会(IEEE-SA)是IEEE内部的一个全球标准化机构,由个人和公司成员以及其他组织的技术人员组成,他们开发基于共识的电工技术、电子、信息和通信技术标准,影响广泛的市场基础。IEEE标准协会提供了一个标准程序,服务于工业、政府和公众的全球需求。在标准化领域,IEEE标准协会是唯一能够代表IEEE的机构。达成的谅解条款如下。

(1)目标。美国电气与电子工程师协会和印度标准局具有直接或间接执行和促进国际标准化的共同目标。

美国电气与电子工程师协会和印度标准局在获得关于另一个组织的活动的知识方面有着互惠的目标和兴趣,这可能有助于在共同感兴趣的项目上进行合作。美国电气与电子工程师协会和印度标准局指出,有必要构建和加强它们之间的关系,并促进更密切的合作。因此,双方制定本谅解备忘录的目标有:①鼓励两个组织之间的交流;②促进各方对标准发展活动的知识共享;③在可能的情况下促进彼此技术小组之间的联系和其他合作。

(2)具体的合作活动。①双方同意每季度就各自的具体标准制定活动交流信息。②双方可以不时地交换在共同感兴趣的领域的其他信息。③从美国电气与电子工程师协会和印度标准局指定一名联络员。④根据规定,IEEE 将在 IEEE-SA 网站上提供一个使用 BIS 标志的 BIS 网站的电子链接。⑤根据规定,BIS 将在 BIS 网站上提供一个使用 IEEE 的标志 IEEE-SA 网站的电子链接。⑥美国电气与电子工程师协会和印度标准局可以确定并将领导人/利益相关者/技术专家与每个其他技术委员会联系起来,并帮助促进与制定标准有关的技术事项的信息交流。⑦每一方仅为本谅解备忘录明确授权的特定和有限目的授予另一方使用该方商标和徽标的权利,但每一方有权在另一方使用该商标和徽标前 30 天以书面形式予以审查和批准。⑧本谅解备忘录不是对财政资源的承诺;任何缔约方之间的单独谈判、记录和承诺,应在实际支出之前做出;除另有协议和书面文件外,每一缔约方将负责其参加本谅解备忘录的相关费用。

(3)其他目标和活动。

(4)任期。

(5)终止。

(6)谅解备忘录的内容。缔约双方约定并同意,该谅解备忘录构成缔约双方之间的完整协议,并取代有关此主题的所有先前的协议,并且除非经缔约双方签署的书面文件或经其正式授权,否则不得修改。

(7)争端解决。本谅解备忘录的适用,解释或执行中可能出现的任何分歧或争议,应通过双方之间的谈判和协商友好解决。

(8)可分割性。该谅解备忘录的条款和条件是可分割的。如果根据任何法律规则,本谅解备忘录的任何条件被视为非法或不可执行,则所有其他条款将继续有效。此外,被认定为非法或不可执行的条款和条件应尽可能地根据双方的意图继续有效。

(9)不可抗力。由于完全无法控制的行为而导致的任何延误或失败,任何一方均不负责。

(10)独立承包商。缔约双方之间的关系应是独立的缔约双方之间的关系,本谅解备忘录中的任何内容均不得解释为构成任何一方作为雇员、代理或另一方成员。在不限制上述规定的前提下,任何一方均无权以任何方式为另一方采取行动或约束另一方,做出陈述或保证或代表另一方执行协议,或以任何方式负责另一方的作为或不作为。

(11)第三方受益人。本谅解备忘录的任何内容,无论是明示或暗示的,均无意于根据本谅解备忘录或由于本谅解备忘录而向除本谅解备忘录的当事方及其各自的继承人和受让人以外的任何人授予任何权利或救济。

(12)契约副本。本谅解备忘录可以在一个或多个对应文件中执行,每一个这样的对应文件应被视为正本,但全部应构成一个相同的文件,并应通过传真或数字副本发送签名页,被视为等同于原件。

(13)工作、任务分配。未经另一方事先书面同意,任何一方均不得转让本谅解备忘录或本协议项下的任何权利、责任或义务。

(14)保障。BIS 应就可能针对 IEEE 主张的所有索赔、损失、费用、支出(包括律师费和专家证人的费用)和判决进行赔偿,并使 IEEE 及其官员、董事、代理和雇员免受损害:①BIS 对 IEEE 标准进行的任何修改或翻译;②任何第三方基于或违反本谅解备忘录中包含的担保的任何索赔而导致的。

(15)机密性、宣传性。本谅解备忘录的条款是严格保密的,BIS 不得将其披露给任何第三方,除非:①经 IEEE 明确的书面同意;②为执行谅解备忘录的条款所必需;③遵守任何适用的法律或法院命令。

(16)注意事项。所有通知应以书面形式,通过挂号或挂号信发送,要求的回执通过航空快递寄达双方指定地址,并在邮寄时生效。或者也可以接受传真发送或带有收据确认的特快邮件发送。

附录 A——采纳协议,包括:A1 目标,A2 IEEE 承诺,A3 BIS 承诺,A4 IEEE 标准作为印度国家标准的采纳和翻译,A5 一般条款和条件。

2.印度标准局与欧盟相关机构签订标准化合作备忘录

2017 年,印度标准局与欧洲标准化委员会(CEN)签订题为"印度在全国范围内采用欧洲标准化委员会 EN 115:1995+A1:1998+A2:2004 自动扶梯和乘客输送机的建造和安装安全规则"的谅解备忘录。具体规定条款如下。

就印度标准局使用 EN 115:1995＋A1:1998＋A2:2004 自动扶梯和乘客输送机的建造和安装安全规则(以下简称 EN)内容的条款和条件达成一致,以便在充分尊重欧洲标准化委员会版权保护规则的情况下制定和分发三项国家标准。

关于欧洲标准化委员会和印度标准局之间执行本谅解备忘录的具体条件如下。

(1)印度标准局应确保每一项国家标准的前言(包括确认案文)包含经欧洲标准化委员会许可转载的欧洲标准 EN 的摘录。

(2)印度标准局应通知欧洲标准化委员会使用本谅解备忘录规定的欧洲标准 EN 内容来发布国家标准,并将国家标准的电子副本发送给对外关系主管。

(3)印度标准局在其任何广告或其他宣传材料或活动中,或在其与客户或公众的任何通信中,不应代表根据本谅解备忘录制定的国家标准。

(4)在印度领土内,根据本谅解备忘录制定的国家标准的销售不受限制。

(5)根据本谅解备忘录制定的国家标准的分配不受任何版税的限制。

(6)印度标准局应准确记录根据本谅解备忘录制定的国家标准的所有销售情况,并使用欧洲标准化委员会为此提供的模板,每季度向欧洲标准化委员会提交一份书面声明。

(7)印度标准局不得使用任何类似欧洲标准化委员会商标或商号的商标,以免造成混淆或欺骗。

(8)欧洲标准化委员会和印度标准局承认,本谅解备忘录自签署之日起为期两年。

(9)经欧洲标准化委员会和印度标准局相互同意,可对本谅解备忘录做出修正。

(10)双方应友好地解决因解释或执行本谅解备忘录而产生或与之有关的任何问题。

(11)任何一方至少提前一个月通知另一方,可随时终止本谅解备忘录。

第 5 章

印度标准化战略的标准化管理

与立法体系类似,印度标准化的行政管理方面也存在着一个复杂而庞大的体系架构。印度标准局被印度中央政府赋予了管理标准化事务的核心地位,其职权包含了标准制定、认证及与国际组织对接等多个方面,在每一个方面都有着重要的表现。但在这些标准化运作的具体方面,印度标准局并非唯一的管理机构,每一个领域都是多个机构并存的局面,尤其是在标准制定方面,各机构之间存在着复杂的关系。本章将首先聚焦印度标准局,具体分析其架构及管理标准化事务的方式,而后分别详述标准制定、认证、涉外对接这 3 个方面存在的标准化行政管理的架构体系。

5.1　印度标准局及其标准化管理

如前所述,印度标准局是印度政府在中央政府层级设立的管理标准化活动的核心机构。尽管难以覆盖全国的全部标准化活动,印度标准局对于标准化的管理还是涉及印度标准化战略的核心领域。

5.1.1　印度标准局的管理架构与基本权限

印度标准局设立在印度消费者事务、食品和公共分配部之下,在行政上归该部管辖。其中,印度标准局局长(President)由消费者事务、食品和公共分配部的内阁联邦部长(Union Minister)兼任,副局长(Vice President)则由该部的部长(Minister)兼任。在实际运作的过程中,印度标准局则设有总干事(Director General)来负责主管该机构的日常事务。

印度标准局设有董事会,其中总共设立 25 个成员席位,由来自中央和地方邦政府、议会、产业界、科研机构、消费者协会及其他专业团体的代表担任。当前,在印度标准局的董事会中,5 席来自消费者事务、食品和公共分配部,1 席来自商务部,5 席来自哈里亚纳邦、马哈拉施特拉邦等地方邦政府,2 席分别来自上议院和下议院,2 席来自消费者协会或相关利益代表,1 席来自农民协会,3 席来自各工业联合会,2 席来自企业,2 席来自科研部门,另有 1 席为印度标准局总干事,1 席暂且空缺。执行委员会则由总干事及来自其他部门机构的代表共 10 人组成,在人员构成上与董事会成员有一定的重合。

日常事务的执行则由向总干事负责的 6 个平行的咨询委员会完成,包括财务咨询委员会、标准咨询委员会、实验室咨询委员会、认证咨询委员会、消费者政策咨询委员会、计划与发展咨询委员会。不过自《印度标准局法案2016》生效开始,上述咨询委员会的设置有所调整,改为财务咨询委员会、标准咨询委员会、合格评定咨询委员会、检测与校准咨询委员会,以及可由印度标准局根据规定设置的其他咨询委员会。其工作分别代表了印度标准局所涉足的重要领域,并根据需要下设委员会或机构。如标准咨询委员会负责印度标准局的标准制定工作,根据标准制定的具体领域下设14个部门委员会,每个部门委员会中都根据具体情况设置了若干个分别负责各种类产品标准制定的技术委员会;实验室咨询委员会负责检测产品标准的实验室的管理工作,包括直属的8个中央层级的实验室及7个在地方邦设置的实验室,还向其他一些实验室授以准予进行标准检测的认可。此外,印度标准局还下设了国家标准化研究院,负责开展标准化的培训工作。

在地域分布上,印度标准局既包括设在新德里的总部,也包括5个地方办公室。后者包括设在新德里的中央地区办公室,设在加尔各答的东部地区办公室,设在昌迪加尔的北部地区办公室,设在金奈的南部地区办公室,以及设在孟买的西部地区办公室,以具体负责各区域的事务。综上,印度标准局的总体行政架构如图5-1所示。

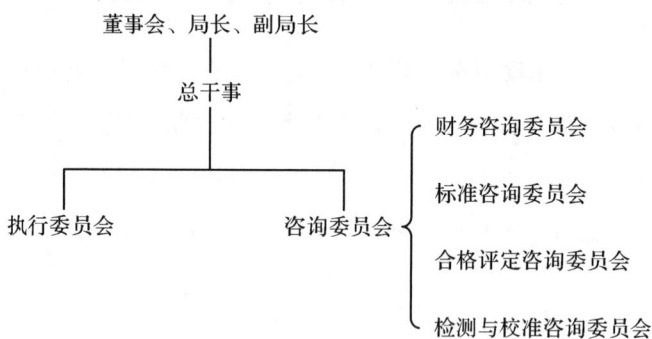

图 5-1 印度标准局的组织架构

5.1.2 印度标准局的标准制定工作

标准制定是印度标准局的一项核心工作。由印度标准局制定出来的标准具有两个基本特点:一是自愿性,不具备强制的效力;二是在印度,只有由印度标准局制定的标准可称为"印度标准",即国家标准。

在"印度标准"制定的组织框架中,印度标准化工作由印度标准局总干事负责统一管理。印度标准局下面主要涉及的部门有 SPPD，标准、政策和计划部，国际关系和技术信息服务部，出版和外语部，信息技术服务部门，以及技术部门。其中技术部门主要负责"印度标准"的制定，目前印度标准局管理的技术部门共计 16 个，包含生产和一般工程部、食品和农业部、化工部、土木工程部、电工部、冶金工程部、电子和信息技术部、石油煤炭及相关产品部、医疗设备和医院规划部、纺织部、机械工程部、交通工程部、水资源部、管理及系统部、服务业部门-I、服务业部门-II。这 16 个领域及其部门技术委员会的设定与印度的产业发展背景密切相关。例如，电子和信息技术、纺织、医疗设备和医院规划都是印度的优势产业，土木工程、交通工程，冶金工程、石油煤炭及相关产品，食品和农业、水资源，这三组产业领域分别与印度政府对于基础设施建设、能源资源以及民生问题的关注息息相关。在人员组成上，这些部门委员会实际上广泛地包括了来自印度国内其他机构部门的标准领域专业人员。以纺织部门委员会为例，据印度标准局的资料，共有 60 个成员，其中包括来自各地的纺织业产业协会、纺织公司、纺织技术科研机构，以及纺织部等政府部门的代表；部门委员会下设的各技术委员会同样如此。

　　印度标准局还为标准制定设定了一套明确的流程。一般来说，印度标准局对于某项标准的制定工作的启动是应相关方的要求，无论是政府其他部门、行业组织、研究机构，还是印度标准局的成员，都可以直接向标准局就某个特定产品领域的标准制定进行提议。在收到提议、进行评估之后，若认为确有必要，则启动标准制定工作。制定流程分为两种基本情况：一是在该领域存在既有的国际标准或其他标准，且标准局认为可以采纳它们作为"印度标准"时，将根据一定程序在既有标准的基础上直接形成"印度标准"；二是当不存在既有的国际标准或其他标准，或这些标准被认为不适合用作"印度标准"时，印度标准局将启动程序制定自己的标准。实际上，在印度标准局看来，即使某些领域已有国际标准可供采纳，也存在着诸多不宜采纳这些标准的情况，包括国际标准与印度的法律规定相冲突、国际标准与印度的实际情况无关等情况。在制定自己的标准的过程中，相应领域的部门委员会、下属的具体产品方向的技术委员会及其在部分情况下根据需要而设立的小组委员会或专门小组负责规范制定进程，在相关方的提议及最终的"印度标准"之间，主要依次形成 3 种文件，包括工作草案（working draft）、初步草案（preliminary draft）以及征求意见稿（wide circulation draft）。其中，印度标准局以后两种文件分别向相关的产业组织等相关方和公众征询意见，在征求意见稿根据相关意见进行修改和审核的基础上，最终形成对外发布的"印度标准"。此外，印度标准局的标准制定在所需时限上共有两个等级,当中央政府或印度标准局认为有必要紧急制定一个标准时,标准制定的时长为 12 个月,而除此以外的所有情况下,标准制定需耗费的时间从 20 个月至 28 个月不等。

截至 2020 年底,累计发布的印度标准数量为 20790 项,其中在研标准为 67 项。这些标准的发布机构分布在印度标准局管理的 16 个技术部门中,其中生产和一般工程部发布的标准数量最多,为 2409 项,其次是食品和农业部门发布的印度标准数量,为 2087 项,化工部发布的印度标准为 1825 项,发布印度标准较少的部门为水资源部 450 项,管理及系统部 383 项,服务业部门-I、服务业部门-II 分别为 51 项和 1 项。

5.1.3 印度标准局的产品认证项目

进行产品认证是印度标准局的职责之一。由其发起的产品认证项目既包括对内的,也包括对外的;既包括自愿的认证项目,也包括强制性的认证项目;既包括具有普遍性的认证项目,也包括针对电子产品、黄金饰品等的特殊认证项目。其中主要的产品认证项目的具体情况如下。

其一,产品认证计划(Product Certification Scheme)。这是一项涵盖了诸多产业领域、产品范围的认证项目,通过认证、获取许可的产品可标上印度标准局的标准标记,以向消费者宣示该产品符合相关的印度标准。截至 2016 年 3 月底,印度标准局颁发的正在生效的产品认证许可共计 31347 项。在这些认证许可中,大部分为制造商自愿申请获得的,印度标准局对于大部分产品的认证不做硬性规定。同时仍有部分产品被划归为需要进行强制性产品认证的范畴,包括水泥、家用电器、奶粉等食品、汽车配件、钢铁等,截至 2020 年底,共计 37 大类 344 种产品,另有数项产品已计划被纳入这一范畴,通过认证后授予 ISI 标志,具体产品如表 5-1 所示。

表 5-1 印度强制认证产品清单

序号	标准号	产品名称
		水泥类产品
1	IS 12330	抗硫硅酸盐水泥
2	IS 12600	低热硅酸盐水泥
3	IS 1489(Part 1)	火山灰硅酸盐水泥(第 1 部分):粉煤灰基
4	IS 1489(Part 2)	火山灰硅酸盐水泥(第 2 部分):煅烧黏土为主

序号	标准号	产品名称
水泥类产品		
5	IS 269	普通硅酸盐水泥
6	IS 3466	砌筑水泥
7	IS 455	矿渣硅酸盐水泥
8	IS 6452	建筑用途高铝水泥
9	IS 6909	特级硫酸盐水泥
10	IS 8041	快硬硅酸盐水泥
11	IS 8042	白色硅酸盐水泥
12	IS 8043	防潮硅酸盐水泥
13	IS 8229	油井水泥
14	IS 16415:2015	复合水泥
15	IS 16993:2018	微细普通硅酸盐水泥
16	IS 15895:2018	高铝耐火水泥
家用电器产品		
17	IS 12640 (Part 1)	家用或相似用途断路器操作剩余电流(第 1 部分):无完整过电流保护断路器(RCCBs)
18	IS 12640 (Part 2)	家用或相似用途断路器操作剩余电流(第 2 部分):有完整过电流保护断路器(RCVOs)
19	IS 13010	交流电电度表,0.5、1 和 2 类
20	IS 13779	交流电静态电能表,1 和 2 类
21	IS 14697	电度表和计时表操作的交流电静态变压器,0.2S 和 0.5S 类
22	IS 15111 (Part 1 & 2)	一般照明用途的自镇流器(第 1 部分):安全要求 &(第 2 部分):性能要求
23	IS 302 (Part 2/Sec 3)	家用和类似用途电器的安全——电熨斗
24	IS 302 (Part 2/Sec 201)	家用和类似用途电器的安全——浸没式电热水器
25	IS 302 (Part 2/Sec 202)	家用和类似用途电器的安全——电炉
26	IS 302 (Part 2/Sec 30)	家用和类似用途电器的安全——室内加热器
27	IS 3854	家用和类似用途的开关
28	IS 418	一般用途电灯泡钨丝(最高 100W)
29	IS 694	工作电压 1100V 及以下聚氯乙烯绝缘电缆

序号	标准号	产品名称
	家用电器产品	
30	IS 8828 修订后为 IS/IEC 60898	电器配件——家用和类似安装电流超载保护断路器
31	IS 9968 (Part 1)	弹性体绝缘电缆(第 1 部分):用于工作电压 1100V 及以下
	电池	
32	IS 8144	多功能干电池
	食品和相关产品	
33	IS 15757	后续配方——补充食品
34	IS 11536	以加工谷物为主的补充食品
35	IS 1165	奶粉
36	IS 1166	炼乳,部分脱脂和全脱脂炼乳
37	IS 12176	超高温甜炼乳
38	IS 13334 (Part 1)	脱脂奶粉,标准等级
39	IS 13334 (Part 2)	脱脂奶粉,特级
40	IS 13428	包装天然矿泉水
41	IS 14433	牛奶蛋白配方的婴儿代乳食品
42	IS 14542	部分脱脂奶粉
43	IS 14543	包装饮用水(包装天然矿泉水除外)
44	IS 1656	牛奶谷物配方的断奶食品
45	IS 3470	己烷,食品级
46	IS 14625	塑料奶瓶
47	IS 5168	玻璃奶瓶
	油压炉	
48	IS 10109	油压炉,偏置燃烧器型
49	IS 2787	多燃烧器油压炉
50	IS 1342	油压炉
	汽车配件	
51	IS 13098	充气轮胎内胎
52	IS 15627	两轮和三轮机动车充气轮胎
53	IS 15633	乘用车充气轮胎斜交和径向帘布层
54	IS 15636	商用车充气轮胎斜交和径向帘布层

续　表

序号	标准号	产品名称
		气缸、阀门和调节阀
55	IS 14899	汽车用液化石油气容器
56	IS 15100	汽车用永久固定液化石油气(LPG)容器多功能阀门组件
57	IS 3196 (Part 4)	超过 5L 水容量低压液化气体的焊接低碳钢气瓶(第 4 部分):腐蚀性毒气体钢瓶
58	IS 3196 (Part 1)	超过 5L 水容量低压液化气体的焊接低碳钢气瓶(第 1 部分):液化石油气(LPG)钢瓶
59	IS 3196 (Part 2)	超过 5L 水容量低压液化气体的焊接低碳钢气瓶(第 2 部分):液化气体(液化石油气除外)钢瓶
60	IS 3224	压缩气体钢瓶(不包括液化石油气钢瓶)的阀门配件
61	IS 3745	用于小型医用气瓶的轭式阀门连接
62	IS 7142	不超过 5L 水容量低压液化气体的焊接低碳钢气瓶
63	IS 7285 (Part 1)	可反复充气的无缝钢气瓶(第 1 部分):正火钢气瓶
64	IS 7285 (Part 2)	可反复充气的无缝钢气瓶(第 2 部分):调质钢气瓶,抗拉强度小于 1100MPa(112 kgf/mm^2)
65	IS 7302	装有通气设备的气瓶阀门配件
66	IS 7312	焊接无缝液化乙炔气瓶
67	IS 8737	水容量大于 5L 的液化石油气钢瓶用阀门配件(第 2 部分):新制造的液化石油气钢瓶用阀门配件
68	IS 8776	用于 5L 水容量及以下的液化石油气(LPG)钢瓶的阀门配件
69	IS 9798	用于液化石油气(LPG)混合物的低压调节器
		医疗设备
70	IS 3055 (Part 1)	体温计(第 1 部分):实心杆型
71	IS 3055 (Part 2)	体温计(第 2 部分):封闭式体温计
72	IS 7620 (Part 1)	医用 X 光诊断设备
		钢铁产品
73	IS 1785 (Part 1):1983	预应力混凝土用普通硬拉钢丝(第 1 部分):冷拉应力消除钢丝
74	IS 1785 (Part 2):1983	预应力混凝土用普通硬拉钢丝(第 2 部分):冷拉钢丝
75	IS 6003:2010	用于预应力混凝土的锯齿形钢丝
76	IS 6006:2014	预应力混凝土无涂层应力消除钢绞线
77	IS 13620:1993	熔接环氧涂层钢筋

续　表

序号	标准号	产品名称
		钢铁产品
78	IS 14268:1995	无涂层应力消除低松弛七股预应力混凝土钢绞线
79	IS 277:2003	镀锌钢板(普通的和波纹的)
80	IS 2002:2009	包括锅炉在内的中高温压力容器用钢板
81	IS 2041:2009	中低温压力容器用钢板
82	IS 2830:2012	一般结构用的铸钢锭、钢坯、大方坯和板坯
83	IS 1786:2008	用于混凝土加固的高强度变形钢筋和钢丝
84	IS 648:2006	冷轧无取向电工钢板及全加工带材(CRNO)
85	IS 3024:2015	晶粒取向电工钢板(CRGO)
86	IS 15391:2003	定向电工钢板和带材
87	IS 2062:2011	热轧中高强度结构钢
88	IS 432 (Part 1):1982	混凝土钢筋用低碳钢和中等强度钢筋和硬拉钢丝(第1部分):低碳钢和中等强度钢筋
89	IS 432 (Part 2):1982	混凝土钢筋用低碳钢和中等强度钢筋和硬拉钢丝(第2部分):硬拉钢丝
90	IS 513 (Part 1):2016	冷轧碳钢板和带材(第1部分):冷成形和拉伸用途
91	IS 513 (Part 2):2016	冷轧碳钢板和带材(第2部分):高强度多相钢
92	IS 1079:2017	热轧碳钢板和带材
93	IS 1875:1992	用于锻件的碳钢方坯、板坯和棒材
94	IS 2879:1998	金属电弧焊焊条用低碳钢
95	IS 3502:2009	钢制网纹板
96	IS 5872:1990	冷轧钢带(箱带)
97	IS 5986:2017	成型和翻边用热轧钢板
98	IS 6240:2008	用于制造低压液化气瓶的热轧钢板(可达6mm)
99	IS 7283:1992	用于生产光亮棒材的热轧棒材和用于工程应用的机加工零件
100	IS 7887:1992	一般工程用软钢线材
101	IS 10748:2004	焊接钢管用热轧钢带
102	IS 11513:2017	冷轧用热轧碳钢带
103	IS 15647:2006	焊接钢管用热轧窄钢带
104	IS 7904:2017	高碳钢线材
105	IS 14246:2013	连续预涂镀锌钢板和卷材

续　表

序号	标准号	产品名称
钢铁产品		
106	IS 15965:2012	预涂铝锌合金金属涂层钢带和薄板(普通)
107	IS 280:2006	一般工程用低碳钢丝
108	IS 1835:1976	用作绳索的圆钢丝
109	IS 3975:1999	电缆铠装用低碳镀锌钢丝
110	IS 4368:1967	一般工程用锻造用合金钢坯、大方坯和板坯
111	IS 4454 (Part 1):2001	机械弹簧用钢丝(第 1 部分):冷拉纯合金钢丝
112	IS 4454 (Part 2):2001	机械弹簧用钢丝(第 2 部分):油淬火和回火钢丝
113	IS 4824:2006	轮胎胎线
114	IS 11169 (Part 1):1984	冷镦/冷挤压用钢(第 1 部分):锻碳和低合金钢
115	IS 11587:1986	耐候钢
116	IS 15103:2002	耐火钢
117	IS 15914:2011	用于焊接气瓶制造的高抗拉强度平轧钢板(可达 6mm)、薄板和钢带
118	IS 15961:2012	热浸镀铝锌合金金属涂层钢带和薄板(普通)
119	IS 15962:2012	建筑用结构钢,增强抗震性能
120	IS 6527:1995	不锈钢线材
121	IS 6528:1995	不锈钢丝
122	IS 6603:2001	不锈钢条和扁钢
123	IS 5522:2014	器皿用不锈钢板条
124	IS 6911:2017	不锈钢板、薄板和带材
125	IS 15997:2012	厨具和厨房用低镍奥氏体不锈钢薄板和带材
126	IS 1110:1990	硅铁
127	IS 4409:1973	镍铁
128	IS 1029:1970	热轧钢带(捞砂)
129	IS 2385:1977	冷轧马口铁板和冷轧黑钢板用热轧低碳钢薄板和卷形带钢
130	IS 3039:1988	船体建造用结构钢
131	IS 9550 2001	光亮钢筋
132	IS 3748:1990	工具和模具钢
133	IS 5517:1993	淬火和回火用钢

序号	标准号	产品名称
钢铁产品		
134	IS 7291:1981	高速工具钢
135	IS 7494:1981	内燃机阀门用钢
136	IS 12146:1987	压力容器用碳锰钢锻件
137	IS 16585:2016	磁性材料,单个材料规范,半加工状态下交付的铁基非晶带材
138	IS 2831:2012	重新轧制成结构钢(普通质量)用碳素钢铸锭、钢坯、大方坯和板坯
139	IS 1148:2009	钢铆钉钢筋(中高强度)结构用
140	IS 1673:1984	软钢线材,冷镦工艺
141	IS 1812:1982	木螺钉制造用碳素钢丝
142	IS 2507:1975	弹簧用冷轧钢带
143	IS 2255:1977	机械螺钉制造用低碳钢线材(冷镦工艺)
144	IS 3195:1992	制造蜗壳和螺旋弹簧用钢(铁路车辆用)
145	IS 3431:1982	汽车悬架用蜗壳弹簧、螺旋弹簧和叠层弹簧制造用钢
146	IS 3885(Part 1):1992	制造叠层弹簧用钢(铁路车辆)(第 1 部分):扁钢
147	IS 3885(Part 2):1992	制造叠层弹簧用钢(铁路车辆)(第 2 部分):肋钢和槽钢
148	IS 4223:1975	伞骨用钢丝
149	IS 4224:1972	订书钉、别针和夹子用钢丝
150	IS 4397:1999	滚珠和滚子轴承保持架/保持架用冷轧碳钢带
151	IS 4398:1994	制造球、滚子和轴承套圈用碳铬钢
152	IS 6902:1973	轮辐用钢丝
153	IS 6967:1973	电焊接圆链用钢
154	IS 7226:1974	一般工程用中、高碳和低合金钢冷轧带钢
155	IS 7557:1982	用于制造冷锻铆钉的钢丝(不超过 20mm)
156	IS 8052:2006	一般工程用弹簧、铆钉和螺钉生产用钢锭、钢坯和大方坯
157	IS 8951:2001	生产高碳钢线材用铸钢锭、钢坯和大方坯
158	IS 8952:1995	一般工程用低碳钢线材生产的钢锭、钢坯和大方坯
159	IS 9476:1980	碳钢剃须刀刀片用冷轧钢带
160	IS 9962:1981	针用钢丝
161	IS 12367:1988	焊接钢管制造用冷轧碳钢带材/卷材

续　表

序号	标准号	产品名称
		钢铁产品
162	IS 14331:1995	高温螺栓用钢
163	IS 14491:1997	冷成形用低碳高强度冷轧钢板和卷材
164	IS 14650:1999	轧制用碳素钢铸锭、方坯、大方坯和板坯
165	IS 4882:1979	轴承工业用铆钉用低碳钢丝
166	IS 2090:1983	预应力混凝土用高强度钢筋
167	IS 6529:1996	用于锻造的不锈钢坯料和厚板
168	IS 9294:1979	刀片用冷轧不锈钢带
169	IS 10631:1983	不锈钢焊条芯线
170	IS 10632（Part 2）:1983	电气用非磁性不锈钢（第 2 部分）:绑线的特殊要求
171	IS 10632（Part 3）:1983	电气用非磁性不锈钢（第 3 部分）:薄板、带材和板材的特殊要求
172	IS 11169（Part 2）:1989	冷镦/冷挤压用钢（第 2 部分）:不锈钢
173	IS 5651:1987	气动工具用钢
174	IS 9516:1980	耐热钢
175	IS 11952:1986	活塞销用钢（活塞销）
176	IS 12045:1987	用于电阻金属加热元件的合金
177	IS 14652:1999	18％镍马氏体时效钢筋和棒材
178	IS 1566:1982	混凝土钢筋用硬拉钢丝织物
179	IS 5489:1975	轴承工业用渗碳钢
180	IS 11946:1987	软磁铁条
181	IS 11947:1987	软磁铁棒、扁钢和型材
182	IS 963:1958	飞机用铬钼钢棒
183	IS 4454（Part 4）:2001	机械弹簧用钢丝（第 4 部分）:不锈钢钢丝
184	IS 1993:2018	冷还原的电解锡板
185	IS 12591:2018	冷还原电解镀铬/氧化铬涂层钢
186	IS 412:1975	一般用途的膨胀金属钢板
187	IS 2100:1970	锅炉用钢坯、棒材和型钢
188	IS 2589:1975	家具弹簧用硬拉钢丝
189	IS 3298:1981	造船用软钢铆钉
190	IS 4072:1975	弹簧垫圈用钢

序号	标准号	产品名称
		钢铁产品
191	IS 8510 (Part 2):1977	电枢和转子捆扎用钢丝(第 2 部分):磁性捆扎钢丝
192	IS 8510 (Part 3):1977	电枢和转子捆扎用镀锡钢丝(第 3 部分):非磁性捆扎钢丝
193	IS 8563:1977	用于制造开口销的半圆软钢线
194	IS 8564:1977	辐条接头用钢丝
195	IS 8565:1977	综丝
196	IS 8566:1977	芦苇用钢丝
197	IS 8917:1978	镀锌板
198	IS 9442:1980	用于制造农业圆盘的热轧钢板和带材
199	IS 9485:1980	用于搪瓷的冷轧和热轧碳钢板
200	IS 10794:1984	开口销用低碳钢丝
201	IS 12262:1988	弹簧垫圈用梯形钢丝
202	IS 12313:1988	热浸钨涂覆碳钢板
203	IS 15911:2010	结构钢(普通质量)
204	IS 4430:1979	模具钢材
205	IS 4431:1978	碳和碳锰易切削钢
206	IS 4432:1988	表面硬化钢
207	IS 5518:1996	落锻用模锻块用钢
208	IS 8748:1978	锻造/轧制 CTC 段
209	IS 12145:1987	压力容器用调质和回火合金钢锻件
210	IS 13387:1992	用于金属成形的工具钢锻件
211	IS 14698:1999	用于制造军用弹壳和防弹用碳素钢和低合金钢坯、大方坯、板坯和棒材
212	IS/ISO 11951:2016	冷还原铁皮制品——黑板
213	IS 3930:1994	火焰淬火和感应淬火钢
214	IS 5478:1969	恒温金属板和带材
215	IS 13352:1992	用连铸坯、钢坯和板坯生产的锻件用坯料
216	IS 16644:2018	预应力混凝土用应力消除低松弛钢丝
217	IS 17404:2020	电镀锌热轧和冷轧碳钢板和带钢
218	IS 8329:2000	用于水、气和污水的离心铸造(旋压)球墨铸铁压力管道

续　表

序号	标准号	产品名称
		钢铁产品
219	IS 9523:2000	水、气、污水压力管道用球墨铸铁配件
220	IS 1161:2014	结构用钢管
221	IS 1239 (Part 1):2014	钢管和其他锻钢配件(第 1 部分):钢管
222	IS 4270:2001	用于水井的钢管(直径不超过 200mm)
223	IS 9139:1979	铸造厂用可锻铸铁丸和磨砂
		变压器
224	IS 1180 (Part 1)	户外型油浸式配电变压器达 2500kVA 及以下,33kV(第 1 部分):矿物油浸式
		电机
225	IS 12615	节能感应电动机:三相鼠笼式
		电容
226	IS 2993	交流电动机电容器
227	IS 13340	额定电压 650V 及以下交流电源系统用自愈型电力电容器
228	IS 13585 (Part 1)	额定电压 1000V 及以下交流系统用非自愈型并联电力电容器(第 1 部分):通用性能、试验和额定安全要求安装和使用指南
		化学品和化肥
229	IS 252:2013	烧碱
230	IS 10116:2015	硼酸
231	IS 15573	聚氯化铝
232	IS 695	醋酸
233	IS 2833	苯胺
234	IS 517	甲醇
235	IS 5158:1987	邻苯二甲酸酐
236	IS 8058:2018	吡啶
237	IS 16113:2013	γ-甲基吡啶
238	IS 16112:2013	β-甲基吡啶
239	IS 12084:2018	吗啉
240	IS 297:2001	硫化钠
241	IS 7129:1992	无水碳酸钾
242	IS 170:2004	丙酮

序号	标准号	产品名称
化学品和化肥		
243	IS 4581:1978	三氯化磷,纯分析试剂
244	IS 11744:1986	五氯化磷
245	IS 11657:1986	氯氧化磷
246	IS 2080:1980	稳定的过氧化氢
247	IS 3205:1984	沉淀碳酸钡
248	IS 12928:1990	陶瓷和玻璃工业用沉淀钡
249	IS 4505:2015	甲醛次硫酸氢钠
250	IS 6100:1984	无水三聚磷酸钠
251	IS 14709:1999	n-丙烯酸丁酯
252	IS 336:1973	乙醚
253	IS 5295:1985	乙二醇
254	IS 537:2011	甲苯
255	IS 15030:2001	对苯二甲酸
256	IS 14707:1999	丙烯酸甲酯
257	IS 14708:1999	丙烯酸乙酯
258	IS 12345:1988	醋酸乙烯酯单体
259	IS 4105:2020	苯乙烯(乙烯基苯)
260	IS 5149:2020	顺丁烯二酸酐
261	IS 12540:1988	氰乙烯
厨房电器		
262	IS 302 (Part 2/ Section 14)	手持搅拌器
263	IS 4250	家用电动食品搅拌机(榨汁机、研磨机)和离心式榨汁机
配合石油气使用的家用热水器		
264	IS 15558	配合液化石油气使用的即时家用热水器
空调及其相关部件、气密压缩机和温度传感控制		
265	IS 1391 (Part 1):2017	室内空调器规范(第 1 部分):单元式空调器
266	IS 1391 (Part 2):2018	房间空调器规范(第 2 部分):分体式空调器
267	IS 8148:2018	管道和包装空调

续　表

序号	标准号	产品名称
空调及其相关部件、气密压缩机和温度传感控制		
268	IS 11329:2018	用于室内空调的翅片式热交换器
269	IS 10617:2018	密封的压缩机
插头、插座和交流直连式静电预付费电表		
270	IS 1293:2005	额定电压 250V 及以下和额定电流 16A 及以下的插头及插座
271	IS 15884:2010	交流直连式静态预付电表(Class1 和 Class2)
与液化石油气一起使用的家用煤气炉		
272	IS 4246:2002	与液化石油气一起使用的家用煤气炉
透明浮法玻璃		
273	IS 14900:2018	透明浮法玻璃
家用压力锅		
274	IS 2347:2017	家用压力锅
电缆		
275	IS 1554(Part 1):1988	工作电压 1100V 及以下聚氯乙烯绝缘(重型)电缆(第 1 部分)
276	IS 1554(Part 2):1988	工作电压为 3.3~11kV 的聚氯乙烯绝缘(重型)电缆(第 2 部分)
277	IS 7098 (Part 1):1988	工作电压 1100V 及以下交联聚乙烯绝缘聚氯乙烯护套电缆(第 1 部分)
278	IS 7098 (Part 2):2011	工作电压 3.3~33kV 及以下的交联聚乙烯绝缘热塑性塑料护套电缆(第 2 部分)
279	IS 7098 (Part 3):1993	工作电压为 66~220kV 的交联聚乙烯绝缘热塑性塑料护套电缆(第 3 部分)
280	IS 14255:1995	天线束状电缆工作电压 1100V 及以下
281	IS 9968 (Part 2):2002	工作电压为 3.3~33kV 的弹性体绝缘电缆(第 2 部分)
282	IS 8784:1987	热电偶补偿电缆
283	IS 9857:1990	焊接电缆
284	IS 14494:2019	矿山用弹性体绝缘软电缆
285	IS 2593:1984	矿用帽灯用柔性电缆
286	IS 5950:1984	发射电缆(竖井以外使用)
287	IS 17048:2018	工作电压 1100V 及以下的无卤阻燃(HFFR)电缆

序号	标准号	产品名称
液化石油气胶管		
288	IS9573(Part 1):2017	液化石油气(LPG)用胶管(第1部分):工业应用
289	IS9573(Part 2):2017	液化石油气(LPG)用胶管(第2部分):国内和商业应用
铝箔		
290	IS 15392	食品包装用铝及铝合金裸箔
玩具		
291	IS 9873 (Part 1):2018	玩具安全(第1部分):机械和物理性能的安全
	IS 9873 (Part 2):2017	玩具安全(第2部分):易燃
	IS 9873 (Part 3):2017	玩具安全(第3部分):某些元素的迁移
	IS 9873 (Part 4):2017	玩具安全(第4部分):秋千、滑梯和类似的活动玩具,供家庭室内和室外使用
	IS 9873 (Part 7):2017	玩具安全(第7部分):手指涂料的要求和试验方法
	IS 9873 (Part 9):2017	玩具安全(第9部分):玩具和儿童产品中的若干邻苯二甲酸酯酯类
	IS 15644:2006	电动玩具的安全性
透明平板玻璃		
292	IS 2835:1987	透明平板玻璃
安全玻璃		
293	IS 2553 (Part 1):2018	安全玻璃(第1部分):建筑、建筑物和一般用途
294	IS 2553 (Part 2):2019	安全玻璃(第2部分):道路运输
编织袋		
295	IS 14887:2014	纺织品高密度聚乙烯(HDPE)/聚丙烯(PP)编织袋,用于包装50kg的粮食
296	IS 16208:2015	纺织品高密度聚乙烯(HDPE)/聚丙烯(PP)编织袋,用于包装10kg、15kg、20kg、25kg和30kg的粮食
297	IS 14968:2015	纺织品高密度聚乙烯(HDPE)/聚丙烯(PP)编织袋,用于包装50kg/25kg糖
298	IS 14252:2015	纺织品填充砂用高密度聚乙烯(HDPE)/聚丙烯(PP)编织袋
蝶阀		
299	IS 13095:1991	通用蝶阀

续　表

序号	标准号	产品名称
	自行车反射镜	
300	IS/ISO 6742 (Part 2):2015	后向反射装置
	纸	
301	IS 14490:2018	普通复印机纸
	牲畜饲料	
302	IS 2052:2009	牛用复合饲料
	汽车轮辋部件	
303	IS:16192（Part 1）	二轮和三轮汽车用轮辋（第 1 部分）:轻合金轮辋
304	IS:16192（Part 2）	二轮和三轮汽车用轮辋（第 2 部分）:钣金轮辋
305	IS:16192（Part 3）	二轮和三轮汽车用轮辋（第 3 部分）:轮辋
306	IS:9436	客车车轮
307	IS:9438	卡车和客车车轮或轮辋
	鞋类	
308	IS 5557:2004	工业和防护橡胶膝踝靴
309	IS 5557（Part2）:2018	所有的橡胶胶靴和短靴
310	IS 5676:1995	模压实心橡胶鞋底和鞋跟
311	IS 6664:1992	用于鞋底和鞋跟的微孔橡胶板
312	IS 6719:1972	实心 PVC 鞋底和鞋跟
313	IS 6721:1972	PVC 凉鞋
314	IS 10702:1992	橡胶沙滩鞋
315	IS 11544:1986	橡胶拖鞋
316	IS 12254:1993	聚氯乙烯工业靴
317	IS 13893:1994	半刚性的聚氨酯鞋底
318	IS 13995:1995	无衬里模压橡胶靴
319	IS 16645:2018	模压塑料鞋,一般工业用内衬或无内衬聚氨酯靴
320	IS 16994:2018	男女用鞋,用于市政清理工作
321	IS 1989（Part 1）:1986	矿工用皮革安全靴和鞋
322	IS 1989（Part 2）:1986	用于重金属工业的皮革安全靴和鞋
323	IS 3735:1996	帆布鞋橡胶底

序号	标准号	产品名称
鞋类		
324	IS 3736:1995	帆布靴橡胶底
325	IS 3976:2018	矿工用橡胶帆布安全靴
326	IS 11226:1993	直接模压橡胶底的皮革安全鞋
327	IS 14544:1998	直接模压聚氯乙烯(PVC)鞋底的皮革安全鞋
328	IS 15844:2010	运动鞋
329	IS 17012:2018	PU 橡胶底高踝战术靴
330	IS 17037:2018	防暴鞋
331	IS 17043:2018	德比鞋
332	IS 15298 (Part 2):2016	个人防护设备(第 2 部分):安全鞋
333	IS 15298 (Part 3):2019	个人防护设备(第 3 部分):劳保鞋
334	IS 15298 (Part 4):2017	个人防护设备(第 4 部分):职业鞋
冲压工具冲头		
335	IS 4296 (Part 1):2016	冲压工具(第 1 部分):60 度锥形头直柄圆冲头
336	IS 4296 (Part 2):2016	冲压工具(第 2 部分):圆柱头直柄或缩柄冲头
337	IS 4296 (Part 3):2016	冲压工具(第 3 部分):60 度锥形头和小柄圆冲头
两轮摩托车驾驶员用头盔		
338	IS 4151:2015	两轮摩托车驾驶员用头盔
制冷电器		
339	IS 1476 (Part 1):2000	家用制冷器具(第 1 部分)
340	IS 7872:2018	冷冻机
341	IS 15750:2006	家用自动除霜制冷设备
离心铸造铁管		
342	IS 1536:2001	用于水、气、污水的离心铸造(旋压)铁压力管道
343	IS 3989:2009	离心铸造(旋压)铁管套土壤、废弃物、通风及雨水管道、配件及附件
344	IS 15905:2011	无轮毂离心铸造(旋压)铁管、配件和附件——水龙头系列

其二,强制登记计划(Compulsory Registration Scheme)。该计划专门针对电子和信息技术产品。在印度通信和信息技术部多次颁布的两道有关"电子和信息技术产品(关于强制登记的要求)行政令"的基础上,截至 2020 年底,共计 63 项电子和信息技术

产品被纳入这一强制登记计划的范畴。该计划既涉及本土制造商,也覆盖进口,63 种电子产品的销售和进口一律以获得许可为前提,如表 5-2 所示。2018 年,新能源和可再生能源部颁布了强制注册计划下的太阳能光伏系统、器件和组件清单,将 5 类光伏产品列入强制注册范围,如表 5-3 所示。2020 年,重工业和公共企业部颁布了前置注册计划下低压开关设备和控制设备清单,将 8 类低压电器产品列入印度强制注册范畴,如表 5-4 所示。

表 5-2　电子和信息技术产品强制注册清单

标准号	标准名称	产品分类
IS 616	音频、视频和类似 电子设备安全要求	电子游戏(视频)
		32 英寸及以上的等离子/液晶/LED 电视
		内置输入功率 200W 及以上扩音器的光碟机
		输入功率 2000W 及以上的扩音器
		输入功率 200W 及以上的电子音乐系统
		用于音频、视频及类似电子设备的电源适配器
		32 英寸以下的等离子/液晶/LED 电视
		无线耳机
		输入功率低于 200W 的电子音乐系统
		除等离子/液晶/LED 电视以外的电视
		无线话筒
		摄像机
		摄像头(成品)
		智能音箱(带显示器或不带显示器)
		蓝牙音箱
IS 13252 (Part 1)	信息技术设备 安全通用要求	笔记本电脑/笔记本/平板电脑
		视觉显示单元,屏幕尺寸 32 英寸及以上的视频监视器
		打印机、绘图仪
		扫描仪

续　表

标准号	标准名称	产品分类
IS 13252 （Part 1）	信息技术设备 安全通用要求	无线键盘
		电话应答机
		机顶盒
		自动数据处理机
		IT 设备的电源适配器
		移动电话
		收银机
		终端售卖机
		复印机/打印机
		智能读卡器
		邮件加工机/邮资机/邮资盖印机
		护照阅读器
		用于便携式应用的充电宝
		视觉显示单元,屏幕尺寸可达 32 英寸的视频监视器
		闭路电视摄像头/闭路电视记录
		USB 驱动的条形码阅读器、条形码扫描仪、虹膜扫描仪、光学指纹扫描仪
		智能手表
		键盘
		自动取款机
		USB 类型的外部硬盘驱动器
		USB 型外置固态存储设备（256GB 以上容量）
		独立开关电源（SMPS）,输出电压 48V（max）
		数字照相机
IS 302 （Part 2:25）	家用和类似用途电器的安全（第 2 部分）:特殊要求第 25 节微波炉	微波炉
IS 302 （Part 2:26）	家用和类似用途电器的安全（第 2 部分）:特殊要求第 26 节时钟	带有电源的电子钟

续　表

标准号	标准名称	产品分类
IS 302 (Part 2/Section 6)	家用和类似用途电器的安全（第 2 部分）：特殊要求第 6 节炉灶、灶具、烤炉和类似用途电器	电磁炉
IS 302 (Part 2/Section 15)	家用和类似用途电器的安全（第 2 部分）：特殊要求第 15 节加热液体用电器	电饭煲
IS 302 (Part 1：2008)	家用和类似用途电器的安全（第 1 部分）：一般要求	家用和类似用途电器的适配器
IS 16242 (Part 1)	UPS 通用及安全要求	额定≤5kVA 的 UPS/变换器
		额定≤10kVA 的 UPS/逆变器
IS 15885 (Part 2/Sec 13)	灯具控制装置的安全（第 2 部分）：特殊要求第 13 节 LED 模块的直流或交流电子控制装置	LED 模块的直流或交流电子控制装置
IS 16046	二次电池和含有碱性或其他非酸性电解质的电池，便携式密封二次电池和用其制成的便携式电池：安全要求	密封二次电池/电池含有碱性或其他非酸性电解质，用于便携式应用
IS 16102 (Part 1)	一般照明设备用自镇流器 LED 灯（第 1 部分）：安全要求	用于一般照明的自镇流器 LED 灯
IS 16103 (Part 1：2012)	一般照明用 LED 模组（第 1 部分）：安全要求	用于一般照明的独立 LED 模块
IS 10322 (Part 5/Sec 1)	灯具（第 5 部分）：特殊要求第 1 节固定通用灯具	固定通用 LED 灯具
IS 10322 (Part 5/Sec 2)	灯具（第 5 部分）：特殊要求第 2 节嵌入式灯具	嵌入式 LED 灯具
IS 10322 (Part 5/Sec 3)	灯具（第 5 部分）：特殊要求第 3 节道路和街道照明灯具	用于道路和街道照明的 LED 灯具
IS 10322 (Part 5/Sec 5)	灯具（第 5 部分）：特殊要求第 5 节泛光灯	LED 泛光灯
IS 10322 (Part 5/Sec 6)	灯具（第 5 部分）：特殊要求第 6 节手提灯	LED 手提灯
IS 10322 (Part 5/Sec 7)	灯具（第 5 部分）：特殊要求第 7 节灯串	LED 灯串
IS 10322 (Part 5/Sec 8)	灯具（第 5 部分）：特殊要求第 8 节应急照明用灯具	应急照明 LED 灯具
IS 10322 (Part 5/Sec 9)	灯具（第 5 部分）：特殊要求第 9 节绳索灯	灯饰链（绳灯）

标准号	标准名称	产品分类
IS 60669-2-1:2008	家用和类似用途固定电气装置用开关的标准	LED产品调光器
IS 16333(Part 3)	移动电话(第3部分):印度语言支持移动电话的特殊要求	移动电话

表 5-3　太阳能光伏系统、设备和组件强制注册清单

标准号	标准名称	产品分类
IS 14286 IS/IEC 61730-1 IS/IEC 61730-2	晶体硅地面光伏组件设计资格和型号批准	晶体硅地面光伏组件(基于硅片的)
IS 16077 IS/IEC 61730-1 IS/IEC 61730-2	薄膜地面光伏组件设计资质和型号认证	薄膜地面光伏组件(a-Si、CiGs和CdTe)
IS 16221 (Part 2)	光伏发电系统用电力变换器的安全(第2部分):逆变器的特殊要求	用于光伏发电系统的电力转换器
IS 16169	电力互联光伏逆变器防孤岛措施试验规程	公用事业互联光伏逆变器
IS 16270	太阳能光伏电池用二次电池一般要求和试验方法	蓄电池

表 5-4　低压开关设备和控制设备强制注册清单

标准号	标准名称	产品分类
IS/IEC 60947 (Part 2:2016)	低压开关设备和控制设备(第2部分):断路器(第一次修订)	低压开关设备和控制设备:断路器
IS/IEC 60947 (Part 3:2012)	低压开关设备和控制设备(第3部分):开关、隔离器、开关隔离器和熔断器组合装置	低压开关设备和控制设备:开关、隔离器、开关隔离器和熔断器组合装置
IS/IEC 60947 (Part 4:Sec 1:2012)	低压开关设备和控制设备(第4部分):接触器和电机启动器第1节机电接触器和电机启动器	低压开关设备和控制设备:机电接触器和电机启动器
IS/IEC 60947 (Part 4:Sec 2:2014)	低压开关设备和控制设备(第4部分):接触器和电机启动器第2节交流半导体电机控制器和启动器(第一次修订)	低压开关设备和控制设备:交流半导体电机控制器和启动器

续　表

标准号	标准名称	产品分类
IS/IEC 60947 (Part 4:Sec 3:2014)	低压开关设备和控制设备(第 4 部分):接触器和电机启动器第 3 节交流电非电机负载用半导体电机控制器和接触器(第二次修订)	低压开关设备和控制设备:非电机负载用交流半导体电机控制器和接触器
IS/IEC 60947 (Part 5:Sec 1:2009)	低压开关设备和控制设备(第 5 部分):控制电路设备和开关第 1 章机电控制电路器件(第一次修订)	低压开关设备和控制设备:机电控制电路设备
IS/IEC 60947 (Part 5: Sec 2:2007)	低压开关设备和控制设备(第 5 部分):控制电路设备和开关元件第 2 节接近开关	低压开关设备和控制设备:接近开关
IS/IEC 60947 (Part 5:Sec 5:2016)	低压开关设备和控制设备(第 5 部分):控制电路设备和开关元件第 5 节具有机械闭锁功能的电气紧急停止装置	低压开关设备和控制设备:具有机械闭锁功能的电气紧急停止装置

其三,金银首饰纯度认证计划(Hallmarking Scheme of Gold/Silver Jewellery)。在印度,金银首饰尤其是黄金首饰的制造与售卖是一项非常庞大的产业,因此为该产业特别设定认证计划也成了印度标准化和产品认证网络的一个特色。一直以来,该计划是金银首饰的生产商自愿向印度标准局申请认证和许可的,到 2016 年 3 月底,已有 15887 项正在生效的许可被授予了相关生产商。370 家受到印度标准局认可的试金及纯度认证中心成为执行这一认证计划的重要主体,从事金银首饰的纯度检验和认证工作。此外,值得注意的是,印度政府一直以来都希望能赋予该项目以强制性,但一直未能在实际执行中得到落实。截至 2021 年 7 月,印度标准局授予牌照的黄金精炼厂共计 41 家。

其四,外国生产商认证计划(Foreign Manufacturers Certification Scheme)。该计划自 2000 年开始启动,专门针对寻求向印度市场出口商品的外国制造商。仅有电子和信息技术产品的认证属于前述强制登记计划,进口中其他各种产品的认证皆属于该计划。在自愿性和强制性的基础上,该计划与产品认证计划重合,即大部分为自愿认证,而产品认证计划中强制认证的产品范畴同样适用于外国生产商认证计划。印度标准局的总部设有外国生产商认证部以负责此事。截至 2021 年 7 月底,印度标准局在这一计划之下颁发的正在生效的许可共计 1007 项,涉及 51 个国家的生产商和 112 项"印度标准"。

其五,生态标志计划(Eco Mark Scheme)。该计划涵盖特定的产品范围,包括肥皂与洗涤剂、涂料、纸张、塑料、化妆品、纺织品、电池、木材代用品、推进剂及喷雾、食品、电子电气产品、包装材料、润滑油、药品、食品防腐剂与添加剂、农药、皮革。可以看出,该计划主要针对家庭日用品,意在促进这些产品达到环境标准。

其六,管理体系认证(Management System Certification)。在这个框架下,印度标准局开展了许多具体的管理体系认证项目,如质量管理体系认证计划(Quality Management System Certification Scheme)、环境管理体系认证计划(Environmental Management System Certification Scheme)、职业健康与安全管理体系认证计划(Occupational Health and Safety Management System Certification Scheme)、危害分析与关键控制点计划(Hazards Analysis and Critical Control Point Scheme)、食品安全管理体系认证计划(Food Safety Management System Certification Scheme)、服务质量管理体系认证计划(Service Quality Management System Certification Scheme)、能源管理体系认证计划(Energy Management System Certification Scheme)等共计 24 项管理体系认证计划。这些管理体系认证涉及工业生产、环境、职业保护等诸多领域。

5.2 印度标准制定机构的体系

5.2.1 印度公共部门的标准制定体系

除了印度标准局之外,在印度中央政府的各部门中还存在着诸多参与标准制定的机构。印度标准局与这些部门一起,构成了一个复杂的公共部门标准制定体系。

就所制定的标准涉及的产业领域而言,印度标准局与其他各部门之间的关系基本划分为两类。如前所述,印度标准局在 16 个领域成立了技术委员会,在这些领域制定标准,而其他参与标准制定的政府机构,既有在其以外的产业领域进行标准制定的,也有涉足这 16 个产业领域中的某一个领域的。具体而言,在标准制定的产业领域上与印度标准局各自独立的机构主要有农业部市场与检查局(Directorate of Marketing and Inspection)、中央药品标准控制组织(Central Drugs Standard Control)、国防部标准化局(Directorate of Standardization, Ministry of Defence)等。与此同时,与印度标准局的标准制定领域出现重复的其他政府部门也不在少数。例如,印度标准局和印度食品安全与标准局(FSSAI)同时进行食品标准的制定工作;印度纺织部(Ministry of Textiles)和印度标准局一样从事纺织品方面的标准制定;在石油产品领域,除了印度标准局以外,还有石油与爆炸物安全组织(Petroleum and Explosives Safety Organisation)、石油工业安全局(Oil Industry Safety Directorate)、石油与天然气管理局(Petroleum and Natural Gas Regulatory Board)等多个部门参与其中。

此外,印度标准局与其他部门之间在制定的标准性质上存在着一个重要的区别。前者仅制定自愿性质的标准。不过这并不代表印度标准局或其他部门在特定的认证项目中不可以将自愿性质的"印度标准"用于强制认证,即印度标准局在制定"印度标准"之时仅为其赋予了自愿标准的性质,而在实施的过程中,中央政府可以在一些具体的产品领域发布规定,即前述强制认证的行政令,从而使这些特定的"印度标准"从自愿性质转化为强制性的标准,并由这些认证所涉及的具体领域的其他政府部门来保证强制性的执行。其他一些从事标准制定的政府部门则可以在制定的时候直接为其标准赋予强制的性质。

在此基础上,印度标准局与其他所涉领域相重合的部门之间的具体关系值得进一步考察。以印度标准局和印度食品安全与标准局对食品标准的制定和实施为例。两者在制定的标准所涉的具体食品产品种类上存在着大幅度的重合,这从两者分别所设的技术委员会可以明显看出(见表5-5)。在同一种产品上,两者所制定的标准既有相互差异的情况,也有相一致的情况出现。2016年底,印度食品安全与标准局曾发布公告称,将采纳印度标准局制定的有关食品添加剂的46种标准,即证明了这一点。

表 5-5　印度标准局食品领域的技术委员会、印度食品安全与标准局科学委员会

印度标准局食品领域的技术委员会	印度食品安全与标准局科学委员会
糖工业	糖果、糖、甜味剂及蜂蜜
食品添加剂	食品添加剂、香料、食品加工助剂
兴奋剂食品	功能性食品、营养品、饮食疗法产品
水果、蔬菜及相关产品	水果、蔬菜及其制品
鱼、渔业及水产养殖	鱼及渔业产品
油及油料种子	油及脂肪
饮料与饮用水	水及饮料

续　表

印度标准局食品领域的技术委员会	印度食品安全与标准局科学委员会
粮食谷物、相关产品及其他农业生产	谷类、豆类及其制品
乳制品及设备	牛奶及奶制品
即食食品及专项产品	转基因生物与食品
烟草及烟草产品	
香料和调味品	

注：印度标准局食品领域的技术委员会及印度食品安全与标准局所设技术委员会/科学委员会都不止这些，此表中仅选取了其中直接围绕食品标准开展的委员会。

　　印度标准局所制定的食品标准是自愿性质的，而《印度食品安全与标准法案2006》则规定，所有的生产商都需要根据印度食品安全与标准局制定的标准进行注册。因此，一般而言，即使两者在相同食品品种上制定的标准存在一定的差异，但在市场进行认证的过程中，由于上述性质的不同，不会在认证上相互冲突。不过值得注意的是，在印度标准局的产品认证项目中，已根据"食品安全与标准，有关销售的禁止与限制规定"等而将部分食品纳入了强制认证的范畴，认证中采用的标准则是印度标准局就这些产品制定的相应的"印度标准"。这些产品包括奶粉、包装饮用水等，它们往往也在印度食品安全与标准局的登记计划之中。因此，在包装饮用水等领域，印度生产商实际上需要分别获取印度标准局和印度食品安全与标准局颁发的许可，且根据规定，在获取印度标准局的许可之前，生产商无法获取印度食品安全与标准局的许可，即后者实际上是以前者为前提的。不过，出于执行力度不足等原因，在印度市场的饮用水等领域，两个机构所颁发的许可并非总是同时为生产商所拥有。例如2016年的一项统计，在饮用水包装方面，在所有获得许可的生产部门中，仅有约1/4的部门同时拥有两个机构的许可，其余的生产部门都仅拥有印度标准局颁发的许可。

5.2.2　印度标准制定中的公私关系

　　除了上述从事标准制定的各政府部门机构以外，同时也有私人部门参与到标准制定之中，构成印度的"私有标准"（private standard）。在印度部分规模较大的产业领域，私人部门在标准制定中颇为活跃。与政府部门标准制定机构的体系不同，印度私人部门的标准制定尚未有一个统一的管理框架，情况因各产业领域而异。它们的标准制定活动与政府部门并非截然分开，两种行为体之间往往有着千丝万缕的关联，甚至时常合作开展标准制定工作,印度政府近年来也试图创造机会促进标准制定中的公私协调关系。其中,信息通信技术领域是印度私人部门参与标准制定以及公私合作的一个典型领域。公私合作的详细内容已经在 4.1.1 信息通信技术领域的标准化部分进行了阐述,此处不再赘述。

5.3 印度标准认证机构的体系

印度标准局发起的数项标准认证计划是印度标准认证体系中的重要组成部分,但非其全部内容。除印度标准局外,还有为数众多的机构组织发起了自己的标准认证项目,它们在涉及的领域上各不相同,在性质和服务对象等方面有同有异,综合起来构成了印度标准认证的体系。

5.3.1 重要的标准认证机构

与印度标准局一样,在印度,对于政府层面的其他一些参与标准认证的机构而言,标准认证往往只是其数项工作和任务中的一种。而在下述机构中,标准认证在其工作中占有重要的分量,相关项目在印度的标准认证体系中也占据引人注目的地位。

其一,印度质量委员会(Quality Council of India)。印度质量委员会是于 1996 年由印度政府与印度产业团体联合成立的一个机构,以在印度各领域推进质量标准的宣传、采用和遵守为己任。为实现这一目标,开展质量认证被纳入了其工作内容。质量委员会下设两类机构,一类是"质量促进"(Quality Promotion)机构,另一类则是审定委员会(Accreditation Boards)。后者共有 4 个具体机构,即国家认证机构审定委员会(National Accreditation Board for Certification Bodies)、国家检测校准实验室审定委员会(National Accreditation Board for Testing and Calibration Laboratories)、国家教育培训审定委员会(National Accreditation Board for Education and Training)、国家医院与保健机构审定委员会(National Accreditation Board for Hospitals and Healthcare Providers)。

其中,国家认证机构审定委员会专门负责全国范围内向参与产品认证的机构颁发许可,国家检测校准实验室审定委员会审核的对象则是参与认证检测的实验室。它们尽管基本上不直接发起产品或服务的认证项目,却与印度全国范围内的认证体系有着非常密切的关联。

印度质量委员会或者直接开展质量认证项目,或者与下设的审定委员会或其他部门的相关机构共同启动质量认证项目。前者如瑜伽专业人员自愿认证项目(Scheme for Voluntary Certification of Yoga Professionals)、印度良好农业实践认证计划(India Good Agriculture Practices Certification Scheme),后者包括与国家认证机构审定委员会共同开展的印度医疗服务认证计划(Indian Certification for Medical Services Scheme)、与国家药用植物委员会(National Medicinal Plants Board)共同启动的药用植物生产自愿认证计划(Voluntary Certification Scheme for Medicinal Plant Produce)。此外,国家教育培训审定委员会和国家医院与保健机构审定委员会更是直接开展针对教育机构或医疗机构的认证项目,以对符合标准、具有资格的教育或医疗机构颁发许可。

其二,印度出口检验委员会(Export Inspection Council of India)。印度出口检验委员会是印度政府在《出口(质量管控与检测)法案1963》的指导之下成立的机构,旨在通过质量控制与检测及其他相关措施促进印度出口的质量。因此,该委员会开展产品认证的服务对象主要是印度国内的出口商。

印度出口检验委员会直接或通过其下属的5个出口检测机构(Export Inspection Agencies),针对出口商的不同需求开展多种类型的质量认证。例如在一些特惠关税项目之下,印度出口检验委员会向相应的出口商颁发原产地认证,这些特惠关税项目有普遍优惠制、南亚区域合作联盟特惠贸易协定、南亚自由贸易区等。另有一些针对特定出口市场的特定商品的检验和认证,如适用于向欧盟国家出口鱼类产品的出口商的健康认证(health certificate)、适用于向欧盟国家出口印度香米的出口商的真实性认证(certificates of authenticity)等。通过在食品加工部门中应用依据国际标准的食品安全管理系统(food safety management systems),印度出口检验委员会向符合相关标准的食品产品颁发质量认证;此外,该委员会根据ISO相关标准,向国内其他一些检验机构和实验室颁发参与出口检验的许可,从而将更多的机构纳入出口检验认证的范畴。

为了促进印度出口,印度出口检验委员会还实施了一些配套措施。例如,印度出口检验委员会对外与其他国家的一些对应机构签订了互认协议,从而印度出口商在该委员会处获得的质量认证在相应的出口市场也能受到认可。前述向鱼类产品及印度香米发放认证许可即是印度与欧盟达成协议的结果。这类协议在中印之间同样存在,具体包括印度出口检验委员会与中国国家质量监督检验检疫总局(现为国家市场监督管理总局)于 2006 年签订的有关铁矿石检验的合作协议,双方于 2013 年签订的有关饲料和饲料添加剂的协议,以及于 2015 年签订的有关印度菜籽粕的协议。

其三,标准化、检测与质量认证理事会(Standardization, Testing and Quality Certification,STQC)。该机构隶属于印度电子与信息技术部,专门负责电子与信息技术产品和服务的标准化及相关事务。实际上,在标准制定方面,标准化、检测与质量认证理事会除了制定电子政务标准之外,基本上只是通过派出人员参与技术委员会等方式辅助和协调印度标准局在电子与信息技术领域的标准制定工作,开展自己的认证项目则是其工作任务中的一个重要方面。STQC 的认证项目主要有基于 IEC 标准的产品安全认证、网站质量认证、智能卡检测与认证、生物计量装置检测与认证、软件及系统认证等产品认证项目,以及信息技术及电子商务领域的 ISO 9001 质量管理体系认证、ISO 27001 信息安全管理体系认证、ISO 20000-1 信息技术服务管理认证等管理系统认证项目。

与出口检验委员会一样,STQC 通过签订一些国际协议,使其认证项目得到国际认可。由该理事会发起的印度通用准则认证计划(Indian Common Criteria Certification Scheme)正是其中重要的一种。STQC 加入了通用准则认可安排(Common Criteria Recognition Arrangements),这意味着印度成了该安排中的一个认证授权国,由 STQC 根据通信准则标准对信息技术产品进行的保证等级 4 级的认证同样为该安排内的其他成员国所认可。

5.3.2　印度各认证项目的分类

综合前述机构包括印度标准局,可以看出,印度存在着种类繁多的产品和服务认证项目,发起这些项目的机构与制定标准的机构并不完全重合,存在着一定的出入,而各类认证项目具体服务的对象和目的也各不相同。因此,各类认证项目在所采用的标准依据的来源上往往有所分别,主要可分为如下两大类。

其一,认证项目及其所采用的标准源自同一机构。印度标准局、农业部市场与检查局等一些机构既是制定标准的主体,也是产品认证的主体,这些机构发起的认证项目基本上采用的是本机构所制定的标准。如印度标准局的产品认证项目等即以其所制定的

"印度标准"为依据。值得说明的是,如前所述,印度标准局的产品认证计划和外国生产商认证计划的一部分以及强制登记计划,都属于强制认证的范畴。在这一部分中,强制认证所采纳的标准仍然是印度标准局所制定的"印度标准",印度标准局所公布的每一项属于强制认证的产品种类,都附有一个"印度标准"的编号。这与"印度标准"的自愿性质并不相违背。在这些强制认证之中,印度标准局像其他自愿认证项目一样,向相关产业领域申请认证的制造商颁发许可,在流程上并无不同;同时,认证的强制性由其他政府机构来确保。如前所述,将某一特定产品纳入强制认证的范畴,是由具体的行政令规定的,行政令也规定了其他政府机构在确保特定产品的制造商全部进行认证上的权利。如2012年的钢铁部行政令规定,"有关当局"有权要求任何参与规定品种的钢铁产品制造、储存或销售的行为体提供必要信息,并检查任何生产中的或用于销售的钢铁产品,而"有关当局"具体包括钢铁部的高级官员、各邦政府区域产业中心的主管、各邦政府负责产业领域的主管。也就是说,印度标准局制定的"印度标准"仍然是自愿性质的标准,但这并不妨碍印度中央政府根据相关考虑将其中特定的标准用于强制认证,而印度标准局尽管是一些强制认证项目的发起者,但这些项目在执行过程中的强制性并不由印度标准局来保证。在这些强制认证项目中,印度标准局与中央政府及其各部各有分工、密切配合。

此外,农业部市场与检查局是印度农产品分级标准的制定者,其所发起的农产品分级认证项目(AGMARK)就是根据自身制定的农产品分级标准。AGMARK 在性质上基本是自愿认证,其中只有两项产品属于强制认证的范畴,分别是食用混合植物油和人造奶油,这两项产品的强制认证是由《食品安全与标准法案2006》规定的,其他自愿认证则归属于《农业生产(分级与标志)法案1937》的管辖范畴,不过它们采用的都是市场与检查局制定的标准。

其二,认证项目的进行采用的是国内外其他机构制定的标准。在印度,有一些发起认证项目的机构自身并不制定同领域的标准,因而其认证项目或采用国内其他机构制定的标准,或依据国际标准,等等。例如,能源效率局发起的一项名为"标准及标签计划"(Standards & Labeling Scheme)的认证项目就综合地采用了这两种来源的标准。该项目启动于2006年,旨在对国内市场上的一些家用电器等种类的产品的能耗进行分级认证,从而令消费者在购买时对产品的能耗情况有确切的认知。如前述诸多认证项目,这一计划同时包含了自愿认证和强制认证,目前,纳入强制认证范畴的产品共有8

种,包括了空调、无霜冰箱、彩电等,而对于每一种产品的能耗分级认证,中央政府在咨询了能源效率局之后都发布了关于认证的规定。根据这些规定,空调、无霜冰箱等产品的制造商必须在以相应的"印度标准"为依据进行认证和根据国际标准进行认证之间进行选择,即"印度标准"及国际标准同时适用于这些产品的强制认证,如空调能耗分级认证的标准包括了"印度标准"IS 1391:1992 及国际标准 ISO 9000;彩电等产品的强制认证规定则直接标明了以相应的国际标准为依据,如该项目中彩电的能耗分级认证标准为 IEC 62087:2008 及 IEC 62301:2011。由此可以看出,"印度标准"并不仅仅适用于印度标准局自身的产品认证项目,也往往被国内其他一些认证机构应用,且对于部分领域的印度市场而言,"印度标准"与国际标准具有相同的效力,是可以在其间进行选择、同时使用的标准。

5.4　印度标准化的涉外对接体系

对于 ISO 等标准化领域的非政府国际组织或其他涉及标准化事务的国际机构,不同的印度机构是印度政府对接和参与不同国际组织的代表。概言之,印度标准局是印度政府对接国际组织的一个最重要的机构,是 ISO 和 IEC 的唯一代表,也是印度参与WTO/TBT-SPS 通报咨询的代表;印度电信部等是参与国际电信联盟的代表,属于食品安全与标准局的印度国家食品法典委员会(National Codex Committee, NCC-INDIA)则代表印度参与国际食品法典委员会(CAC)。

印度标准局是印度政府在 ISO 中的官方代表。为了更好地参与到 ISO 中,印度标准局设立了国家对口委员会,以承担 ISO 合格评定委员会(CASCO)及消费者政策委员会(COPOLCO)的对口工作。印度自 1911 年即开始参与到 IEC 之中,为了参与 IEC,印度设有印度 IEC 国家委员会(IEC National Committee of India)。1949 年,印度标准协会接替了先前的工程师协会(Institution of Engineers)开始负责印度 IEC 国家委员会。当前由印度标准局担任该委员会的秘书处。印度标准局当前在 ISO 和 IEC 中的具体参与情况已在前文中做了详细说明,这里不再赘述。

电信产业的标准化是国际电信联盟的重要工作任务之一。印度自 1869 年开始参与国际电报联盟(国际电信联盟的前身)。目前共有 16 个印度的政府或非政府机构参与到国际电信联盟中。其中,隶属于印度通信部(Ministry of Communications)的电信部(Department of Telecommunications)以及印度电信管理局(Telecom Regulatory Authority of India)是政府部门,前者作为印度中央政府在国际电信联盟中的代理,更是享有管理者的身份。此外,印度著名的私人性标准制定机构电信标准发展协会(Telecommunications Standards Development Society)也参与其中,其他机构则是相关的电信领域研究机构、企业或产业团体等。

为了与国际食品法典委员会保持联系并支持其在印度的相关工作,印度食品安全与标准局成立了印度国家食品法典委员会以及印度国家食品法典委员会联络点(National Codex Contact Point,NCCP-INDIA)。具体而言,印度国家食品法典委员会的主要功能和职责包括就国际食品法典委员会正在开展的标准化工作向印度政府提供咨询、派代表参加CAC的会议、制定印度在CAC相关事务中的官方立场等。印度国家食品法典委员会联络点则担当CAC秘书处与NCC-INDIA及其对口委员会之间的联络者,作为NCC-INDIA的秘书处运作,并具体参与到跟踪国际食品标准、开展相关研究等标准化工作中。

此外,作为WTO的成员之一,印度也参与到TBT/SPS通报咨询体系之中。在印度商务部的委任之下,印度标准局担当印度的TBT/SPS通报咨询点,一方面向WTO其他成员国通知有可能影响国际贸易的国内技术规定方面的变动,另一方面向印度国内的利益相关方发布TBT/SPS的通告,并对其中可能影响本国贸易利益的通知进行分析和评估。印度标准局的TBT/SPS通报咨询工作覆盖除食品、动物及其产品、植物以外的所有产业领域,在这些领域,印度向WTO上报的技术规定的通报内容由印度商务部的贸易政策局负责。上述几个特殊领域中则设有各自的通报咨询点。印度食品安全与标准局担当印度食品领域的SPS通报咨询点,并负责该领域上报WTO的通报内容。印度农业与农民福利部中的农业合作与农民福利部门负责的领域是SPS植物保护咨询通报,SPS动物健康咨询通报则由印度农业与农民福利部下设的畜牧业、乳业及渔业部门负责。

第 6 章

印度标准化战略对我国的启示

　　以上各章描述了印度标准化战略的进程和框架，从中可以看出，印度在推进标准化事务方面已取得了不少成果，有着丰富的经验，同时其标准化战略也正处在不断调整的过程之中。中印两国同属新兴经济体，在产业发展和对外贸易的目标与需求方面，中印两国之间存在不少相似之处，同时两国的国情和国家经济政策也不乏差异。在这样的情况下，对于中国而言，印度的一些关键的标准化政策和实践往往既有其正面的示范效应，也存在着反向的借鉴意义，这两个方面都是中国在推进自身的标准化进程时可以考虑的。

6.1　印度标准化的战略方向带来的启示

　　如前所述，印度的标准化战略与其进出口政策、产业发展战略及国家宏观经济进程之间存在着密切的关联。在将标准化作为维护国家进出口和产业发展利益的有效工具方面，印度的相关实践对于中国而言具有借鉴意义，而其中的负面效应也足以引发思考。

　　一方面，标准化可以在促进出口、推动国内产业融入区域产业价值链的过程中发挥重要作用。在雄心勃勃的出口目标及标准问题阻碍出口的认知的推动之下，印度政府从多种渠道出发极力促使标准化的实现。综合前文对于印度标准化实践的种种描述来看，印度政府努力的方向主要有两个方面。一者，在国内机构方面，设立了多个综合性或专门领域的机构，通过它们一方面积极关注国际标准的动向，并在国内予以协调；另一方面通过产品认证的实施及相应的出口优惠条件，促使出口商符合国际标准。除了前述的印度出口检验委员会，标准化、检测与质量认证理事会以外，类似的机构还有农产品及加工食品出口发展管理局（Agricultural and Processed Food Products Export Development Authority）、海产品出口发展管理局（Marine Product Export Development Authority）等。这些机构积极地参与和签订双边或多边的互认协议，拓展印度出口认证的效用。二者，在国际舞台上，印度政府积极参与 ISO 等组织的标准制定活动，明确地设定了担当标准制定者而非标准遵从者的目标，意在使新制定的国际标准更符合印度产业的需求和利益。这两个方面也都应当成为中国在标准化方面予以更多重视的方向。

另一方面,在利用标准化管控进口的过程中,需要在限制低质量产品进入国内市场和防止过度的贸易保护之间进行谨慎的平衡。印度政府近年来愈加密集地发布行政令,将越来越多的产品纳入强制认证的范畴,并覆盖了面向进口商的认证。其具体指向有二:一个是明确宣称的限制质量低劣的产品通过进口的渠道进入印度国内市场,从而维护消费者利益,改善印度市场秩序;另外一个指向则是较少明确表达出来的,对国内相关产业进行保护、避免其受外来具有竞争力的产品的冲击的意图。这使得印度利用标准化管控进口的实践之中,包含有正面的意义,以及通过标准化的相关措施提升国内市场上的产品质量是非常有必要的;然而与此同时,在将标准化当作贸易保护的工具使用时,带来的负面效应则多过正面意义。

具体而言,其一,在以标准化措施保护某些特定本土产业时,往往会对国内的其他产业带来较大的负面影响。例如,在印度政府将不锈钢纳入强制认证的范畴之后,印度国内原先依赖进口其他国家价格较低的不锈钢原材料以生产相关制成品的制造商,则在很大程度上难以为继,在印度标准局设定的不锈钢标准之下,相当一部分国际不锈钢原材料制造商难以通过认证,而这尽管在某种意义上保护了印度本土不锈钢原材料产业,却使得其他相关产业的生产成本大大增加,并由此在国内引发了不少的不满。其二,以标准化之名行贸易保护之实往往会招致其他贸易伙伴的批评,且与以标准化促进出口的目标难以真正协调。例如,美国即指出,印度国内的一些标准制定与国际标准有出入,使得美国部分领域的产品在进入印度市场时遭遇阻碍。这种状况的出现实际上也会加大印度通过协商打开国际市场的难度。其三,包括标准化在内的贸易保护措施尽管可在短期内为印度国内产业发展提供保护,但从长期来看,并不利于印度国内产业竞争力的增强,与其融入区域产业价值链的目标也有所出入。对此,由莫迪政府组建的国家转型委员会(NITI Aayog)明确提出了与国内较为普遍的贸易保护主张不同的声音:"对于印度本土制造商而言,设定'印度标准'将可带来垄断利润。然而如果其他国家并不认可我们的标准,国内市场的规模将难以扩大,从而限制生产水平……在匆匆将'印度标准'强加于市场之前,我们需要的是创造一个有利于商业发展的生态环境,并使市场发展壮大。过早地实施标准将扼杀产业的增长。"

因此,对于未来的印度而言,如何发挥以标准化管控进口的正面效应,而又关注国内产业的长远发展,在出口目标和进口政策之间找到更为恰当的结合点,是其标准化战略进程中面临的一个重大挑战。对于同为新兴经济体的中国而言,实际上也存在着相同的问题,其中,促进国内制定和实施的标准与先进的国际标准协调统一,是中印两国都需进一步努力的方向。

6.2 印度标准化的管理架构带来的启示

从前面几个章节的描述中可以看出,印度在标准化领域的行政和立法管理的架构模式非常复杂,这种复杂之中既有其秩序性所在,也存在着无序之处,这两点对于中国而言都有其借鉴意义。

具体而言,印度标准化领域行政管理架构的基本特点在于,它同时汇聚了集中性和分散性。集中性指的是,印度政府不仅设立了印度标准局这个专门负责标准化事务的核心机构,而且正在尝试加强其在标准化管理中的职权与地位。分散性则是指,印度标准局的存在及其职权的加强并没有取代中央政府中的其他各相关部门机构对于标准化的管理,印度食品安全与标准局等诸多机构仍然在各自的产业领域中从事着标准制定和认证的工作,甚至其职权和影响力也在同步地不断扩大;而且即使是印度标准局,其人员组成上也同时包括了专门人员及来自其他各部的官员。在印度标准局与其他机构部门之间,既有分工关系,也有相互配合的关系。这种分散性架构模式的存在主要源于两种因素:一者,在国土广阔、产业部门众多的印度,印度标准局的人力、物力等资源难以覆盖这一国度中所有的标准化事务,在这种情况下,令各部门在各自领域推进标准化,并协调印度标准局的相关工作,有助于在具体领域推进标准化的专业性,并给标准化管理事务注入尽可能多的资源;二者,一直以来,印度政治中的集权性较为弱势,各部门机构固有的权力难以轻易撼动,即使强势如莫迪,在统一各部权力和行动的能力上也有其限度,这意味着在推进标准化管理的过程中,需要在印度标准局职权的加强与其他部门权力的维系之中保持平衡。

印度的标准化立法体系方面存在着类似的问题。一方面,《印度标准局法案 1986》及 2016 版本是标准化方面的核心法案,为印度标准局的运作,印度标准局框架下的标准制定、认证和许可颁发等提供了基础性的法律依据;另一方面,这前后两部法案毫不涉及具体产业领域的标准化管理问题,对于各领域,无论是标准化管理的具体运作方式,还是对何种产品应当实行强制认证的问题,都需要在其他相应领域的法案中寻求依据,或不断出台新的法律规定,对未曾说明的内容予以法律解释。

从上述经验出发,对于中国而言,印度标准化行政和法律管理的架构模式主要有三点启示。其一,加强标准化管理核心部门的职权与地位势在必行,有助于推进标准化管理。核心部门职权的加强不仅可以更好地推进自身发起的标准制定和认证等项目计划,也有助于开展与其他相关部门之间的协调工作,从而增强全国标准化管理事务的有序性。其二,标准化管理的核心部门与其他相关机构部门、核心法案与其他各领域法案之间恰当的分工与配合有其必要性所在。标准化工作的推进是各产业部门都应当予以充分重视的,与核心部门相比,具体各产业领域的相应部门机构在推进标准化上有其经验、资源等方面的优势,充分发挥这些部门的支持性作用,能够较有效率地推动标准化在各领域的发展。其三,分散性也会导致混乱和冲突,如何加强各部门之间的协调,在分工的基础上构建一个更为简明、顺畅、统一的宏观框架,是亟待解决的问题。在印度,印度标准局与其他相关部门之间各自为政的情况并不少见,这不仅意味着各部门之间在标准化工作上不必要的重叠及资源浪费,也加重了制造商等各方的负担,且不利于与国际市场在标准化方面的对接;同时由于各部门对于标准化的重视力度不同,印度的标准化进程在各产业领域中一定程度上存在着不平衡发展的问题。

6.3　印度标准化的具体措施带来的启示

在推进标准化的具体措施中,印度的实践既有不少创新之处,也在标准执行的市场监管等方面存在着不足,这些都可为我国的标准化进程提供有益借鉴。

印度创新的措施包括对产品认证等实施电子化管理、召开国家标准大会、向国外提供标准化培训项目等。这些措施都有助于扩大国家标准化进程的影响力,以更为高效的方式将标准化的各利益相关方凝聚在一起,促进相互间的沟通和协调。以印度近年来的标准化重点项目为例,无论是智慧城市标准化项目、网络安全的标准化,还是"绿色计划",都采取了紧跟国家经济社会建设进程的策略。这种策略具有两方面的积极效应,一方面,可以极大地发挥标准化对于经济社会发展的促进作用;另一方面,也可借助国家宏观经济项目使标准化进程吸引更多社会关注、获取更多支持。特别是就后一方面而言,在印度,尽管中央政府一直以来对标准化不乏关注,但是这一议题在社会中的影响力颇为有限，而这也限制了标准化的执行效果。与广受关注的国家宏观项目和计划结合起来，印度标准化最新的几个项目，尤其是智慧城市等，近来在媒体上也逐渐成为讨论的对象，即使各方对其实施不乏争议。这有助于在社会中逐渐形塑对于标准化的认同。此外，这些标准化项目往往有一个共同的特点，即在制定标准的过程中将既有的国际标准和国内特殊情况同时纳入考量之中，试图将两者结合起来。这种综合的做法使得其新设定的标准既可在一定程度上受到国际认可，又可对国内的特殊需求有针对性。不过，由于印度的这些项目基本上处于初始阶段，在制定和实施标准的过程中，能否真正将两者结合起来，避免偏重任何一方或两者之间存在的潜在冲突，都有待于在其未来的进展过程中予以进一步观察。对于中国而言，无论是在国家重大经济社会项目和标准化进程之间相互借力，还是在开展标准化项目的过程中融合国际标准与国内需求，都是值得学习的内容。

在标准执行的市场监管方面，印度政府并非没有予以重视。例如，印度标准局法案的更新以加强市场监管为重点，而印度食品安全与标准局等机构也以此为己任。然而迄今为止，效果仍然有限，印度市场上违反标准和许可认证规定的情况屡见不鲜。究其原因，一方面，政府标准化管理机构未能设定严密的市场程序以保障许可认证的合法合规使用，且实施市场监管的人力严重不足，即使消费者对违法使用许可认证的产品和厂商进行举报，也往往难以得到处理；另一方面，产业界，尤其是广大的中小企业尚未形成有关标准的共识和自觉意识，且政府主导的产品认证的程序往往较为复杂，需耗费大量的时间和金钱成本。因此，加强标准执行的市场监管需要从政府管理部门和产业界两方面着手，加大对市场监管的投入，适当简化产品认证的程序并降低费用，培养产业界的标准意识，多管齐下，改善市场标准执行的乱象。

第 7 章

推进中印标准化合作的建议

目前为止,中印两国政府间的标准化合作非常稀少。但实际上,无论是从双边层面还是多边层面而言,两国间在标准化事务上都存在着广阔的合作空间。中印双边贸易逐年攀升,标准问题在双边贸易中占据的分量也逐渐显现,标准合作的不足是阻碍两国贸易关系深入发展的关键因素,而一旦开展合作,标准也有望在双边贸易中发挥重大的推进作用。与此同时,中印两国都有着参与国际多边标准合作的期望,在全球治理中的丰富的合作经验、对于世界贸易的一些相似立场及在区域产业价值链中地位的相近,都为两国在多边平台上开展标准化合作奠定了基础。为此,本章将依次提出对于推进中印两国标准化双边合作及多边合作的建议。

7.1　推进中印标准化双边合作的建议

从印度标准局的对外合作网络可以看出,印度政府并非不愿在双边标准化合作上采取行动,但中印两国间的正式合作迟迟未能启动。要使以印度标准局为代表的政府机构愿意与中国进行对话、开展合作,一方面需要在更高层次的双边政治经济对话中确定标准化合作的必要性与地位,为标准化合作的开展提供足够的政治动力;另一方面需要向印方展示出足够的使其在标准化合作中获取利益的可能性,为标准化合作的开展提供充分的经济动力。

近十年来,尽管中印间贸易的总量迅速增长,但双边贸易仍然存在着较大的甚至日渐显著的问题和障碍。从印度的角度来看,逐年拉大的对华贸易逆差已成为印度对外贸易领域的一个最重大的顾虑。根据印度商务部的统计,在 2007 财年,印度对华贸易逆差约为 162.75 亿美元,而至 2019 财年,该数值飙升至约 512.4 亿美元。在这十几年中,与中国对印出口一路攀升形成鲜明对比的是,印度对华年出口金额基本未有大的变化。这种不平衡性的原因之一固然在于印度制造业发展的不足,在印度看来,另一关键原因则在于其进入中国市场时面对的非关税壁垒,其中即包括了标准问题。有印度学者指出,正是中国在标准等方面的规定阻止了印度在医药、农业、信息技术等领域的对华出口表现与其在这些产业领域中的实力优势相匹配。不仅是印度,中国在对印出口的过程中同样需面对标准方面的挑战。亟欲解决贸易逆差问题的印度正在将标准化作为一个手段,限制中国对印出口。据相关报道显示,就在 2017 年,印度开始对玩具、电子产品、机械等产品种类的进口进行更为严格的质量检查,这被认为正是针对中国所采取的措施,且上述领域均是中国产品在印度市场上占据相当优势的领域。实际上,无论是对印度还是对中国而言,这种极度不平衡的状态以及双方均存在的非关税贸易壁垒问题,限制了双边贸易的可持续性。从上述状况可以看出,标准化方面的协调与合作是两国共同解决双边贸易面临的问题的关键。

双边标准化合作应以当前中印两国间的进出口状况为根据,首先聚焦于少数几个对双边贸易具有重要意义的产业领域,在获取一定的合作经验之后,再向其他领域扩展,以期拓展双边贸易的发展潜力。中短期内首要的合作领域建议为信息通信技术、钢铁及钢铁制品以及医疗产业。信息通信技术是中印双方共同关注的重要产业领域,中国当前是世界上最大的信息通信技术产业出口国,而印度在对外提供信息通信技术服务方面则有着较大的优势。如前所述,信息通信技术是印度标准化的重点领域之一,且印度正在制定网络安全标准,目的包括对中国在印制造及对印出口设置限定条件,同时打开中国市场。因此从中短期来看,当前正是与印度就信息通信技术的标准化问题展开讨论、寻求合作的窗口期。钢铁及钢铁制品是中国对印出口的重要方向,近年来也是在对印贸易中遭遇反倾销调查的重点领域。印度目前已将 38 种钢铁及钢铁制品纳入强制认证的范畴,以此管控该领域的进口。而为了提升达成合作的可能性,印度至为关注的仿制药和医疗器械出口问题可一并纳入双边讨论与合作的范围。

为切实地推进合作,中印两国可以考虑从两个主要方面着手,发挥标准化在改善双边贸易关系中的作用。其一,建立平台,促进中印两国在标准方面的信息互换。在这一方面,美印之间的标准化合作,尤其是美国在印发布美国标准名录,以及建设两国间的标准信息平台等实践,是可供借鉴的有益经验。美印双方的合作正是以促进双边贸易为目的的。就中印的情况而言,同样可以建设一个中英文双语版的专门平台,用以发布两国在各产业领域的具体标准规定,以及在进出口上的检查与认证要求和相关流程。这些信息也可以印制成册,以电子或纸质的形式通过双方的政府部门及产业团体向两国的产业界发送。这种信息互换将有助于两国企业在相互贸易的过程中不因标准问题而无谓地浪费资源,并助力双边贸易的便利化。为了达成这一目标,可以启动谈判,在印度标准局与中国国家标准化管理委员会之间达成包含信息互换等内容的谅解备忘录。这也将有助于促进两国在标准问题上的战略互信。

其二,成立标准化联合工作组,深入发掘双边标准和认证协调的可行性。仅仅是信息互换还难以解决两国贸易在标准问题上面临的诸多矛盾,更需切实采取措施,解决两国在进入对方市场时面临的标准方面的非关税贸易壁垒问题。建立联合工作组,就是要集中力量共同研究对策,并且应当主要从 3 个方面着力。一是加强两国进行标准互认的可能性。尤其是对于两国重点关注的贸易领域,如印度在对华出口中关心的医药、农业领域,以及中国在对印出口过程中遇到诸多问题的钢铁、电子等领域,尽管两国很难在标准的设定上保持一致,但如果能够实现标准互认,将极大地削减非关税贸易壁垒。此种进程还将有助于促进双边贸易的互惠性,为贸易不平衡的解决打下一定基础,从而有助于在一定程度上阻止其中任一方产生利用标准化来进行贸易保护的意图。二是加强国际标准在中印贸易中的适用性。除了标准互认以外,使国际标准通行于两国间贸易,也可以缓解双方在标准领域的矛盾。中印两国近年来在不少产业领域中对于国际标准的认可程度都在逐渐增加,这意味着在双边贸易中适用国际标准的空间也正在逐步扩大。三是降低进口认证的成本的方法。在中印双边贸易中,即使是产品或服务符合对方有关标准的要求,申请质量检查、获取进口许可的过程也将耗费不少时间、人力和金钱上的成本,两国进出口管理部门就更多的产业领域签订质量认证的互认协议或通过协商互相开通认证的快速通道等,将有助于解决这一问题。

7.2 推进中印共同参与标准化多边合作的建议

不仅是双边层面,中印两国的标准化合作在多边层面上也存在着广阔的空间和潜力。一方面,中印在全球治理领域的良好的合作经验为两国在多边平台上就标准化议题进行合作奠定了一定的基础。同为新兴经济体的中印两国在全球贸易治理、金融治理、气候治理等诸多领域有着共同的立场,合作成果颇为丰硕。在此基础上,可以发挥标准化对于巩固中印两国在全球治理中合作的作用。另一方面,仅就标准化领域而言,中印两国在多边平台上的利益和需求也不乏结合点。两国实际上都希望能够参与到国际标准的制定之中,拓展自身在国际标准制定议程中的话语权,也都愿意以标准化为一种手段促进区域经济一体化。由此,中国可以积极推进与印度在标准化领域的多边合作,可供合作的平台既包括 ISO、IEC 等国际性的标准制定机构,也包括其他一些政府间综合性全球治理机制。

在国际性标准制定机构中，中印两国可围绕如下几点开展合作。其一，在印度标准局与中国国家标准化管理委员会之间建立渠道，在一些标准制定过程中进行沟通，协调双方立场，从而探索就具体的国际标准的制定共同发声的可能性。其二，协同组织 ISO 和 IEC 的一些活动，如技术委员会与分委员会的工作组会议等，或在印度组织此类会议时给予一定支持，这将有助于双方在标准制定的议程设置上发掘合作空间。其三，加强在 ISO 的发展中国家行动计划等项目中的合作，例如在该行动计划的框架下，共同发起对于其他发展中国家标准化发展的援助项目，设立中印联合标准化基金，组建派赴其他发展中国家的专家小组，等等。

对于金砖国家等中印共同参与的政府间全球治理机制而言，标准化合作议程尚处于萌芽或未启动的状态，中印共同的行动对于这些平台上标准化多边合作的启动和发展将大有裨益。以金砖国家为例，金砖国家工商理事会正在积极推动金砖国家间的标准化合作，2017 年厦门峰会上通过的成果性文件之一就是《金砖国家工商理事会关于标准监管合作的共同宣言》。然而，对于金砖国家的标准化合作而言，这是远远不够的。金砖国家应当借鉴 IBSA 在这一领域的合作经验，积极推动在各国中央政府的层面上，或是各国核心的标准化管理局的层面上，签订政府间谅解备忘录或其他类型的合作协议，以切实建设各国有关标准化的信息交流机制、各国标准化管理部门间的定期会晤机制、标准化领域的人才交换机制等。然而，目前印度标准局对于这一合作的兴趣和热情尚有欠缺，因此需要中国牵头在更高的层级上与印度对话，从而将印度标准局纳入这一合作体系之中。

附　录

附录 1　印度标准化战略英文原文

Indian National Strategy for Standardization

FOREWORD

The rapid growth of the Indian economy, its size and emerging relevance in global trade, makes it essential to establish a robust "Quality Ecosystem" in India with a harmonized, dynamic, and mature standards framework. This would fuel economic growth and enhance the "Brand India" label. Standards have been widely recognized as catalysts for technical development, industrial growth, the well-being of the society and more recently for convergence of new and emerging technologies. The growing influence of standards and technical regulations, and corresponding conformity assessment procedures on trade and commerce has been recognized worldwide through the Agreements on Technical Barrier to Trade (TBT) and Sanitary and Phytosanitary Measures (SPS) of World Trade Organization (WTO). Countries are accordingly evolving strategies to synergize standardization work with technological, social and economic development at the national level as well as for playing influencing roles in global standardization efforts.

The Indian National Strategy for Standardization (INSS) considers the current state of development across sectors, the existing quality infrastructure and the policy directions in relation to domestic economic developments and for trade in goods and services.

This INSS is the result of a broad consensus arrived over consultations held over a four-year period from 2014 to 2017 through national and regional standards conclaves that attracted wide participation of experts and stakeholders from union and state governments, industry, regulatory bodies, national and overseas standards and conformity assessment bodies, academics, and international fora.

SCOPE AND APPROACH

The INSS addresses four broad pillars of the Quality Ecosystem viz: (1) Standards Development; (2) Conformity Assessment and Accreditation; (3) Technical Regulations and SPS Measures; (4) Awareness and Education. It determines the critical role for each pillar and sets goals thereunder. Each goal is supplemented by a brief description of the background conditions and recommends specific activities that need to be undertaken for its realization. It takes into account the needs and expectations of all stakeholders and accords the interests of MSMEs a high consideration.

THE STRATEGIC INTENTS

The INSS provides direction for India's political and executive leadership on how best to use standardization, technical regulations, quality infrastructure and related activities to advance the interests and well-being of Indians in a global economy. It is based on the following considerations:

◆Positioning standards as a key driver of all economic activities relating to goods and services.

◆Developing a comprehensive ecosystem in India for standards development taking into account the diversity of interests and expertise available.

◆Using standards for providing a level playing field to domestic industry and enhancing the competitiveness of Indian goods and services in domestic and international markets.

◆ Adopting best practices in standardization, conformity assessment and accreditation, and technical regulations, and creating an integrated infrastructure roadmaps, and institutions for their effective management.

◆Playing an active role and taking leadership positions in apex international fora in the related areas.

◆Creating response mechanisms to global developments on standards, technical regulations, and conformity assessment that impact market access of Indian goods and services.

◆Aligning the Strategy with other national policies related to trade and industry, consumers and environment.

IMPLEMENTING THE INSS

The INSS is intended to be a living document with an implementation plan to ensure positive outcomes in each of the identified areas. Some of the recommended tasks in the INSS do not currently fall under the direct jurisdiction of any of the existing organizations. Once approved, an implementation plan would be prepared to identify the agencies, the related activities to be undertaken by them and the time frames.

While some of the goals and the related activities can be accomplished over a shorter period, it is expected that all elements of the strategy can be undertaken and completed over a five-year period (2018—2023).

The implementation plan would be monitored by a high-level committee with quarterly reviews and the results of the monitoring with the achievements will be published on the India Standards Portal.

STANDARDS DEVELOPMENT

Vision: Building a national culture of standards for growth and economic leadership.

Mission: Developing a dynamic relevant and priority driven standards ecosystem that will drive development across sectors, promote competitiveness of Indian products and services, and foster India's eminence among the global leaders in standardization.

Goal 1:Convergence of all standards development activities in India

➤Enhance capability of SDOs for dynamic and faster development of standards, matching with the pace of technology development.

➤Encourage setting up of new SDOs in emerging technology areas with international connects.

➤Adopt the SDOs standards as national standards when required. Set up a recognition scheme for SDOs.

➤Avoid duplications, conflicts, and overlaps.

➤Ensure market relevance of standards produced at all times.

India's standards setting process has been led by the Bureau of Indian Standards (BIS), the national standards body, since 1947. Additionally, to a limited extent, sector-specific standardization work is carried out by more than 25 other bodies including ministries, regulatory bodies, public sector undertakings, technical development agencies, commodity boards, industry and professional bodies, etc. More recently overseas standards development organizations have established offices in India with a view to engaging experts and to support industries using their standards.

While BIS develops Indian standards through its sectional committees under 14 Division Councils representing sectoral interests based on the international code of practice for standards development, the other SDOs essentially rely on dedicated expertise for standards writing and follow their own procedures. A system to recognize or integrate their standards as national standards do not exist presently.

In order to expand the base and enhance the pace of the standards setting activity, it is essential to enhance the capacity and resource base of the existing SDOs and also to encourage the setting up of new SDOs in new and emerging areas and cutting-edge technologies notably digital technologies, sustainable practices, clean energy and smart cities. These SDOs must have connections and working arrangements with the respective international bodies to ensure that there is no gap in the availability of standards to the Indian industry.

The Bureau of Indian Standards shall remain the apex national standards body and in accordance with the mandate of the BIS ACT, 2016 continue to oversee the harmonious development of standardization activities under its own umbrella as well as through memoranda of understanding with the other SDOs. Standards thus developed by other SDOs can be adopted/adapted as national standards as and when required. The adoption of the standards can be in to(or)relevant part(or)modified to suit India's national requirements and priorities. Such arrangements would ensure that there is no duplication conflict, or overlap in the standardization activities of multiple agencies. It shall remain their joint responsibility to ensure at all times that the standards in force are relevant and reflect the state of the technology and industrial practices through appropriate market surveys, environmental scans and comparative studies with the standards being developed worldwide.

A recognition scheme for SDOs shall be established to ensure that they implement the WTO codes of practice and principles for standards development and any relevant decision in this regard by the concerned committee.

Goal 2: Setting up a dynamic mechanism for new standards identification development and their revision

➢Make standardization a key priority area across all sectors.

➢Set up dialogue fora and processes to articulate and prioritize needs for standards development.

➢Create opportunities for Indian business through standards.

➢Focus on critical sectors linked to economic social and sustainable development.

Over the past seventy years, more than 20,000 Indian standards have been developed, almost 50% of which are product standards and the rest are support standards such as test methods, terminology, codes of practices, etc. The subject matter for taking up new standards or for the adoption of international standards is decided by the respective Division Councils and Technical Committees of BIS, while the other SDOs take similar decisions through committees or through executive decisions.

The present system does not present full opportunity or channels for the articulation of needs by all potential standards users. As a consequence, there is widespread use of standards developed by overseas bodies without their adoption or adaptation in India. There are also large gaps where no standards exist for use, especially in the service sector. In several areas, only guidance standards or codes exist but not the related product standards.

There is an urgent need to create fora and processes to articulate and prioritize needs for standards development in different sectors. The best candidates for articulating the needs are the ministries and policy fora under them, the related industry bodies, export promotion bodies, and the commodity boards wherever present. A standards forum needs to be set up in each of these bodies, that would be responsible to set up stakeholder consultations and dialogue fora with businesses and professional bodies and MSMEs to identify the gaps, to collate the needs and to coordinate with BIS and the relevant SDOs for a time-bound development or revision of the required standards.

The prime considerations during stakeholder consultation would be identification of technologies and the markets and industry sectors, as well as state level requirements where standardization can create opportunities for Indian business.

Goal 3: Inclusive participation of all stakeholders in standards development including States and MSMEs

➢Enhance awareness on the role and benefits of standards and conformity assessment practices among businesses, government and civil society including consumes.

➢Creation of States and District fora for standards related activities.

➢Set up funding mechanisms to supplement participation costs.

Effective standards development requires an adequate and continuous participation of all interest groups and subject matter experts. In order to achieve this, it is important to generate widespread interest and awareness in the standards programmes being undertaken by BIS and other SDOs and to attract participation with financial support where funding becomes a restriction, especially for MSMEs, civil society groups including consumer groups, and identified experts.

The Standards Conclaves initiated since 2014 have served as good fora to inform and educate on the relevance of standardization, conformity assessment, and technical regulations. There is a need to continue holding more Standards Conclaves at the Central and State levels on a rotational basis.

Over time, the standards building process should become a layered activity with inputs invited and collated from States and District standards fora. Depending upon the intensity of the industrial or economic activity, the Standards Committees should give representation for direct participation to these fora.

The funding assistance by Central Government for participation in standards development activity should be broad-based and made more liberal, to fund the participation of MSMEs, individual subject matter experts, non-profit bodies and civil society groups both in national as well as international standards committees. State Governments must be encouraged to fund participation in national standards setting in areas of their interest.

Goal 4:Harmonizing standards with international standards
➢Harmonize national standards with international standards where possible.
➢Avoid dissonance of national standards with market driven standards.
➢Special focus and pooling of expertise on converging technologies.

The need for harmonizing Indian standards with international standards for reducing technical barriers to trade and improving market access for Indian products and services cannot be overemphasized. In order to ensure that Indian businesses remain competitive both in domestic and overseas markets, the goods and services must conform to globally accepted standards, subject to national priorities and requirements. While national considerations should be given due accord, these requirements should be minimal and least trade restrictive. BIS and other Indian standards bodies that participate in international standards bodies [such as ISO, IEC, ITU, Codex Alimentarius Commission, World Organization for Animal Health (OIE), and the International Plant Protection Convention (IPCC)] need to undertake a systematic programme for adopting, adapting and aligning relevant Indian standards with the corresponding international standards. The principal focus should be on product/equipment hardware standards with concomitant adoption/adaption of related codes of practices, test methods, and protocol standards.

Apart from standards developed by international standards organizations, it is widely recognized that standards developed by several professional and/or private bodies have a significant market presence and user base. BIS and other SDOs engaged in the respective domains need to ensure that their standards do not present conflicts to industry and businesses and endeavour to avoid dissonance with the private standards, especially where there is some form of compulsion on conformance to the national standards.

The advancement of ICT and digital technologies in all spheres of manufacturing and service domains will be presenting new challenges for standards developers. This will call for convergence of multidisciplinary expertise into more and more standards projects in the future. Participation from industry having access to international developments in these projects would be imminent. Close collaboration among BIS, Telecommunications Engineering Centre (TEC) and Telecommunications Standards Development Society, India (TSDSI) would be required to create the necessary synergies and avoid duplication of work.

Goal 5: Identifying sectors where India could pioneer standardization work
➢Develop pioneering standards in areas of traditional strength.

➢Develop service standards on a time bound programme.

➢Develop innovative and cutting-edge standards based on research Development.

Historically, the Indian standardization efforts has been to follow standards developed by other standards bodies. India offers many sectors with significant commercial potential that have remained outside the scope of standardization. Undertaking pioneering standardization work in these areas would not only unleash their commercial potential but also enhance India's contribution to global standardization efforts. These areas can include the Indian system of alternative medicines Ayurveda, Yoga and Naturopathy, Unani, Siddha and Homoeopathy (AYUSH), Indian food cuisines, Indian traditional arts, Indian crafts, Indian traditional veterinary systems, etc.

With a view to developing cutting-edge standards in spin-off sectors such as space, and life sciences-sectors that can propel Indian businesses as leading suppliers in the global markets-concerted efforts are required to develop innovative standards based on original research and development. Unlike the traditional approach, these standards would be torchbearers and lead commercial development.

Goal 6: Systematic and continuous participation in international and regional standardization work

➢Ensure continuous participation in international standards committees by identified experts through liberal funding support.

➢Take leadership roles in technical committees and governance structures.

➢Play an active role in development of private standards.

Participation in international standards setting projects enables the voicing, consideration and possible inclusion of national priorities and concerns. With increased adoption/adaption of international standards by the global community, these are becoming de-facto universal standards impacting competitive positions. It is therefore essential that they do not contain provisions that place Indian suppliers at a disadvantage and this can be secured only through the continuous participation of experts who understand the technical requirements as well as their impact on trade and commerce.

An essential pre-requisite is the continuity of participation during the lifecycle of the standards project. A firm policy needs to be in place to ensure positive participation in every standards project that is identified to be in India's interest. Each standards body, regulator, and the related Ministry should identify the international fora related to it, and the committees where India's participation on standards setting or compliance mechanisms is essential. For each of these fora, the participation plans should be drawn synchronized with the annual calendar of meetings and suitable technical experts identified for continuous participation. The experts can be drawn from industry, or scientific bodies or in individual capacity. The fund allocation by the government should fully cover the costs for participation in international standards development activity with simplified procedures that allow automatic funding based on agreed principles.

Continuous participation should be gradually translated into taking leadership positions on international standards and project committees as well as governance structures and winning secretarial responsibilities commensurate with India's position as a leading global economy.

Particular attention is required for leadership in international fora dealing with development and review of International Standards, Guidelines and Recommendations (ISGRs) with respect to all aspects of SPS measures, especially in light of increasing use of such measures adopted provisionally.

With the increasing influence of private standards, participation in the related fora should be sought and attended following a similar approach as for international fora. Some SDOs invite participation of individual experts. Accord inationmechanism needs to be developed under the aegis of BIS to ensure that the individual experts participating in these bodies from India are appropriately sensitized about national priorities and needs.

The South Asian Regional Standards Organization (SARSO) was established in 2010 to achieve and enhance coordination and cooperation among SAARC. Member states in the areas of standardization and conformity assessment with the objective to develop harmonized standards for the region, to facilitate intra-regional trade, and to enhance access in the global market for the SAARC Region suppliers. Being the leading economy of the region, India should play an active role in providing leadership both in the development of the standards as well as in trade negotiations for market access of South Asian products and services on behalf of the SAARC member countries.

In addition to the strengthening of SARSO, India may take a lead role in exploring possibilities for further regional cooperation in standards development.

Goal 7: Development of Service Sector Standards

➢Set up a national task force to accelerate service standards development work.

➢Identify service quality gaps, standards required for related infrastructure and occupational skills.

➢Develop fast track national standards based on gap analysis.

➢Take a leadership role in international service standards development work.

The topic of standardization in services is relatively new and is also weakly addressed in trade law instruments. Further, the inherent characteristics of services (ephemeral nature, heterogeneity, services embedded in goods, etc.) make it challenging to standardise them. However, with services accounting for a major share in the Indian and global economy, there is an urgent need to develop standards inservice.

The Government of India has identified twelve Champion Services Sectors (CSS) for focused attention so as to realise the potential of these sectors. These include Information Technology and Information Technology Enabled Services (IT & ITeS),

Medical Value Travel，Transport and Logistics Services，Tourism and Hospitality Services，Accounting and Finance Services，Audio Visua Services，Communication Services，Legal Services Construction and Related Engineering Services，Environmental Services，Financial Services and Education Services.

The interventions in the identified Champion Sectors initiative are based on five pillars aimed at giving an impetus to these sectors. These include inter-alia New Standards，as the role of standards in shaping the export competitiveness of various services sectors is becoming an increasingly important aspect of global trade in services. Accordingly，the departments and concerned ministries have been mandated to take forward the initiative for development of service standards，including both cross-cutting horizontal standards and sector-specific vertical standards in all champion sectors and their adoption in a time-bound and systematic manner through mandatory as well a voluntary routes，so as to enhance competitiveness service delivery quality and consumer welfare.

India has an advantage in skill-based and labour-based services and has the potential to be an outsourcing destination for many of such services. Ministry of Skills Developmentand Entrepreneurship is already working on strengthening the necessary regulatory framework and qualification standards and institutional accreditation process for skill development inexisting as well as emerging services in line with global standards. This area is critical and needs to be quickly brought in place for promoting the export of various services and creating supportive skilling infrastructure for the identified Champion Services Sectors，particularly at/around major hubs/clusters.

Presently in India，many of services sectors are governed by statutory standards or buyer-driven service agreements. In select areas，the existing service standards match global norms，however，in most cases，the standards need to be raised significantly to be at par with world-class standards.

In order to accelerate standardization efforts，a National Task Force should be set up with the mandate to：

a) Identify the service quality gaps in each of the Champion Service Sector with the help of experts having global exposure in the respective sectors.

b) Develop service standards, including both cross cutting horizontal standards and sector-specific vertical standards in all champion sectors to address this gap and to work on their adoption in a time bound and systematic manner through mandatory as well as voluntary route.

c) Identify the standards required for service quality and supporting infrastructure.

d) Determine the occupational roles for the development of the related skills standards and creation of a matching training and personnel certification framework, including foreign language skills.

e) Take on the leadership role in standard setting in areas where India is seen as the world leader like in IT/ITeS, traditional systems of medicine, yoga, etc.

The task force may comprise of representatives from Department of Commerce as the nodal agency, Bureau of Indian Standards, Ministry of Skills/NSDC, nodal ministries/departments of Champion Service Sectors, apex industry bodies, Quality Council of India and invited experts drawn from each sector. The recommendations of the task force should form the inputs for the development of service sector standards by BIS on fast track basis and for providing/upgrading resources by the relevant ministries in line with the standards.

As services sector is the key driver for economies worldwide, especially India, we should take early lead in the development of service sector standards and leverage the efforts to take leadership positions in international standardization work.

Goal 8: Creating an ecosystem to meet the challenges from Private Sustainability Standards

➢Identify all Private Sustainability Standards Programmes that impact exports.

➢Set up national response structures comprising of experts and seek a voice in the programmes standards setting process.

➢Create corresponding schemes wherever needed and develop expert resources to evolve national ecosystem and facilitate easier compliance.

Over the past few years, a new set of standards termed as Private Sustainability Standards (PSS), sometimes known as Voluntary Sustainability Standards (VSS), has become globally popular. These are built on three fundamental pillars namely social progress, economic development and environment & climate, in line with the Sustainable Development Goals (SDGs). While PSS are impacting trade in significant terms, these are outside the purview of the existing WTO regime. They are driven either by buyer consortiums or institutions working on sustainable development. A supplier of the respective commodity has little choice but to conform to the requirements even though the cost burden in many cases is significantly high, especially for small operators in developing countries. Most of the PSS come in a package, combining the standards, the conformity assessment procedures and the auditor qualification norms. There is a recognition that approximately 500 plus PSS are operating world over, with more than 30 impacting the Indian market. Many of the PSS do not have mechanisms for global stakeholder consultations and are not participatory or transparent.

There is a need for a mechanism to respond to the challenges posed by such standards at the national level. An initial task would be to identify all PSS that are impacting exports from India and to set up national response structures comprising of experts that could seek membership or be the voice in the standards setting process.

Some of the global programmes on PSS are based on benchmarking schemes that conform to common criteria and rules. Wherever such opportunity exists, at least one national scheme should be developed and supported to get recognized under the programme. Even where benchmarking is not available, developing corresponding schemes in India would enable the creation of the necessary eco-system that would facilitate easier compliance. There is also a need to create resources in the market which can help upgrade industry for such standards and provide the required level of hand-holding/counselling/training support.

CONFORMITY ASSESSMENT, ACCREDITATION, AND METROLOGY

Vision: Evolving a credible, competent and robust infrastructure for conformity assessment.

Mission: Provide confidence to customers and markets, supplement and provide alternates to regulatory oversight and escalate Indian exports.

Goal 1: Enhance credibility of conformity assessment programmes in domestic and foreign markets

➤Multiple modes of conformity assessment procedures are available.

➤Encourage all conformity assessment bodies to get accredited by national accreditation boards.

➤Set up a market surveillance agency to monitor conformity caims both under regulation and self-declarations.

➤Encourage and recognize voluntary self-regulation mechanisms for delivery of credible conformity assessment services as alternate to regulations.

The opening up of the Indian economy has led to the simultaneous development of a national quality ecosystem that now includes a full complement of conformity assessment schemes comprising third party inspections, product certification, management systems certification, personnel certification, testing and calibration, self-declaration of conformity, etc. Conformity assessment services are being provided by dedicated government appointed organizations, private Conformity Assessment Bodies (CABs), and in some cases, directly by the regulatory bodies. In the same period, the national accreditation boards have enhanced their scopes and consolidated their outreach by accrediting all the major CABs, in some cases, directly by the regulatory bodies. A significant number of CABs however still operate outside the register of the national accreditation boards that present gaps in their oversight and accountability. As self-compliance levels are still below par, it is important that the entire conformity assessment infrastructure operates with demonstrated high integrity to inspire confidence among domestic and overseas buyers and regulators. Concerted efforts should be made through suitable incentives and mandates to bring all conformity assessment operators within the fold of national accreditation.

Presently, there are limited means for post-market testing of the actual level of conformance of the products placed in the market, especially those that are under self-declaration of conformity. As more and more products are expected to be brought under technical regulations, there is a need to establish a national market surveillance programme that would monitor conformity of products placed in the market, whether certified or regulated under self-declaration of conformity. The results of such surveillance would provide valuable inputs for impact assessment and on the effectiveness of the conformity assessment programmes.

While accreditation boards are responsible for ensuring the competence and independence of the CABs, a supplementary means to raise the credibility of conformity assessment is by the creation of self-regulating and self-managed mechanisms that take responsibility for monitoring the integrity and reliability of the conformity assessment programmes and the CABs that operate them. Responsible CABs and laboratories need to develop such voluntary self-regulating mechanisms and create market pressure for all operators to subscribe to them. The development of such credible mechanisms would reduce the need to regulate and allow markets to operate freely. As an alternative, the registration and operations of CABS in India may be regulated by Department of Industrial Policy and Promotion (DIPP) in an appropriate manner.

Goal 2: Secure and enhance global equivalence through mutual recognition agreements in accreditation and sectoral fora across a broad range of goods and services

National accreditation boards, like NABCB and NABL, have gained full membership across a range of multilateral recognition arrangement in accreditation of conformity assessment programmes within the IAF and ILAC umbrella creating greater inroads and access into foreign markets. This enables the meeting of minimum qualifying criteria for market access by certified products and services within the fold of such accreditations. However, in most cases, regulatory requirements of importing countries require additional conformance conditions that need to be met through equivalence recognition arrangements.

These apply in diverse fields including food, pharma, chemicals, toys, medical devices, electrical and IT equipment, telecommunication equipment as well as services such as IT/ITeS, education and skills certification. It is necessary to map the entire spectrum of products and services that need to meet specific conformity assessment requirements and to systematically install and develop facilities and infrastructure that would help achieve equivalence status. As the department responsible for trade, this initiative should be led and coordinated by the Department of Commerce in association with various relevant ministries/regulators through concerned bodies like the Export Inspection Council, BIS, NABCB, NABL and supplemented by the export promotion councils, commodity boards, accreditation boards and the CABS operating in the respective areas.

Goal 3: Promoting Indian products through a "BrandIndia" label for global acceptance
➤Develop a Brand India Label Scheme.
➤Certification to Indian, international and importing country standards.
➤Brand promotion by export promotion organizations, industry bodies, missions, etc.

Promotion of Indian products needs to be backed with a visible and credible Brand India certification label that gives assurance of quality and sustainable practices which is also globally acceptable. Several national certification schemes such as the BIS Product certification (ISI Mark), Ag Mark for agricultural produce, FSSAI Mark for processed foods, Silk Mark for silk products, Star labelling for energy efficiency of appliances are in existence, over some decades. However, in their present forms, they are constrained by legal boundaries in terms of territorial application as well as the criteria for certification which is limited to Indian standards. In order to expand the outreach of the products for a global audience, the Brand India label would need to be significantly large in scale and operated on professional lines at par with leading global certification programmes. This would require a scheme based on international best practices with a provision for benchmarking existing national schemes.

The requirements of this approach would entail the following changes:

a) Certification for exported products should be offered to Indian, international and importing country standards according to the needs.

b) Multiple routes for product, process and service certification should be available or permitted by relevant ISO CASCO Standards.

Once the scheme is in place, it should be actively promoted by export promotion organizations, industry bodies, overseas missions with funding assistance from Government for brand promotion.

Goal 4: Minimize costs of conformity assessment, especially for MSMEs to make them globally competitive

➤Extend the funding assistance for MSMEs to all types.

➤Promote NABCB/NABL accredited conformity results and domestic equivalent approvals for acceptance by overseas regulators and organized foreign buyers.

➤Set up common testing facilities for MSME clusters.

There is an urgent need to ease the compliance burden of MSMEs in meeting regulatory and overseas market access requirements. Both Central and State Governments have been providing funding assistance for securing Quality Management Systems certification based on ISO 9001 and in some sectors, such as food testing, costs to set up laboratories. Since compliances are typically based on testing, inspections and product certification, the financial assistance needs to be extended to all types of conformity assessment within the allocated budgetary provisions.

The compliance burden can be further reduced by negotiating with overseas regulators and buyers to accept domestic testing, inspection and certification results backed by accreditation provided by NABCB and NABL. Where capability approval is a pre-requisite, Export Inspection Council, export promotion boards, and other relevant organizations need to set up robust conformity assessment schemes in line with international models and seek approval of overseas regulators, scheme owners, or organized buyers that would authorize them to issue certificates of conformity on similar lines as already achieved in some sectors.

As costs of testing equipment is a major burden on MSMEs, State Governments need to identify common needs especially where MSMEs are present in clusters and assist in setting up common testing facilities that can be run on a cooperative or private basis. All laboratories set up under this programme need to be provided technical assistance in setting up nationally traceable measuring arrangements and getting accredited by NABL.

Goal 5: Active participation in international organizations dealing with Conformity Assessment

India is a participating member of ISO CASCO. A mirror committee has been set up by BIS with participating members drawn from accreditation boards, industry, SDOs, PSUs. CASCO standards have a deep impact on the manner in which conformity assessment programmes are mandated to run worldwide. However, due to cultural differences, the presence or absence of requirements impact the effective implementation of these standards. It is absolutely essential to participate in all CASCO working groups by the relevant agencies and attend all their meetings to ensure that India's viewpoints are raised and incorporated in the standards developed.

As the accreditation boards are automatic members of IAF/ILAC, they also need to ensure that they participate in all meetings of these apex organizations.

Similar active participation should be ensured by National Physical Laboratory

(NPL), in the BIPM (the International Bureau for Weights and Measures) and in International Organization of Legal Metrology (OIML) by Department of Legal Metrology.

Whenever an opportunity arises, the key objective should be to occupy leadership positions on the technical committees, working groups and governance structures in these apex bodies.

TECHNICAL REGULATIONS AND SPS MEASURES

Vision: Securing the highest degree of protection for the well-being and safety of Indian citizens.

Mission: Ensuring that technical regulations are aimed to achieve legitimate objectives minimal, risk based, least burdensome, and effective in meeting the objectives with least disruption to businesses.

Goal 1: Develop a sound understanding of good regulatory practices and regulatory impact assessment

➤Adopt good regulatory practices and issue policy guidelines for development, implementation, review and revision of Technical Regulations and regulatory impact assessment.

➤Create coordination and understanding among agencies responsible for notifying and ensuring compliance to technical requlations.

➤Conduct regulatory impact assessment for all technical requlations.

As regulations are issued to protect and balance the needs of civil society and its various interest groups, they need to be precisely calibrated to the risks in context with the times, entail minimal cost burden should be easy to comply, and be transparently administered. They must not impede social development and economic growth. Inappropriately applied technical regulations may lead to higher prices of goods and services, poor service quality and lack of product innovation. As regulations have the tendency of losing relevance with time, they need to be regularly recalibrated for effectiveness purpose.

All technical regulations and SPS measures should be based on the principles of good regulatory practices that include a risk based selection of regulatory measures, considerations of regulatory efficiency, i. e. : balance between costs of compliance and administration versus gains effectiveness in compliance; transparency in notification, administration and changes; openness in communications and balancing of interests.

Technical regulations should also be assessed for impact on benefits against costs, economic burden on government, and impacts on the competitiveness of the industry, market openness, small businesses, public sector and potentially affected social groups.

Policy guidelines based on good regulatory practices and regulatory impact assessment need to be established for the development, implementation review, and revision of Technical Regulations.

It is also necessary to create a thorough understanding of the importance of following good regulatory practices and regulatory impacts among ministries, regulatory bodies, state governments, enforcement agencies, conformity assessment bodies and social groups. Civil Services Academies should initiate awareness courses and workshops on the subject.

Goal 2: Separation of institutional roles to increase effectiveness and to avoid potential conflicts of interest

➢Identify gaps between India and global practices on technical regulations notified.

➢Identify gaps between India and global practices on technical requirements (standards/essential requirements) included in technical regulations.

➢Eliminate the gaps through a systematic plan.

Due to historical reasons the roles of line ministries regulatory bodies, and conformity assessment operations, have become centred within the respective ministries/departments. However, these often present conflicts of interest. An institutional and progressive separation of roles and retention of only essential functions to strengthen the effectiveness of the regulatory framework is necessary. In order to achieve this goal, the following separation of roles is required.

1. Ministries

a) For bringing the necessary enabling legislation for a regulatory framework and for establishing the related regulatory body/agency (e. g. FSSAl Act and PNGRB Act).

b) Creating the required policy and enabling rules for the regulatory framework.

2. Regulatory Body

a) Notifying specific regulations/orders under the applicable Act.

b) Setting up the necessary market surveillance (including port control), conformity assessment, and enforcement frameworks including conducting search and seizure operations, prosecuting non-compliant manufacturers/service providers.

3. Accreditation Bodies

To accredit conformity assessment bodies as per international standards and/or requirements prescribed by the regulators.

4. Conformity Assessment Bodies

To provide third party conformity assessment services(certification, inspection, testing) independent of the regulatory body as per requirements prescribed by regulators.

5. Market Surveillance Authorities

a) To conduct pre and post market testing of products intended to be or placed in the market (including port control).

b) Conducting search and seizure operations, prosecuting non-compliant manufacturers/service providers.

Goal 3: Ensure protection in areas that are widely regulated worldwide

The WTO SPS and TBT agreements provide the enabling considerations based on which technical regulations and measures are notified by countries/economic unions. The considerations include human health & safety; animal and plant life, and health; the environment, national security; and prevention of deceptive trade practices through suitable technical regulations/SPS measures. The products and measures are notified on risk-based assessments.

Comparative studies of products and services regulated in India with the majority of other countries have revealed significant gaps both in terms of numbers as well as the technical requirements to which they must conform. The absence of technical regulations and SPS measures adversely impact the protection of consumers, animal, plant and environment from unregulated sub-standard products and services produced or imported into India. Inflow of sub-standard products which are typically cheaper, also impact the competitive position of responsible producers, many of whom have selected voluntary certification. It is also recognized that minimum compliance levels enhance sectoral capabilities and export potential.

It is essential to conduct a gap analysis of areas where technical regulations and SPS measures have been notified by a majority of countries but not done by India. The analysis should be used as the basis to systematically eliminate the gaps while applying the principles of good regulatory practices and impact assessment. For SPS measures, the analysis should include reciprocal arrangements.

A comparative review of standards/essential requirements notified in Indian technical regulation and SPS measures in light of global practices also needs to be undertaken for suitable modifications and upgradation.

Goal 4: Technical Regulations and SPS measures should be based on appropriate standards/essential requirements, and conformity assessment procedures commensurate with attendant risks and market conditions

➤ Technical regulations must specify minimum essential requirements and be based on established standards or technical requirements drawn by experts.

➤Technical regulations must select the less burdensome route of conformity assessment that is capable of covering the risks.

➤ Technical regulations should be finalized only after wide stakeholder consultation.

Technical regulations and SPS measures invariably create a burden of compliance on the producers and suppliers of products and services with attendant costs that are borne throughout the supply chain including the consumer, as well as government in running the enforcement programmes. Regulators therefore need to carefully select the most essential requirements in products, processes or services that would serve the purpose of the regulations. While the adoption of standards professionally developed by standards bodies should be the first preference, regulators must be proactive in the development and revisions of such standards and ensure that their scope is restricted to aspects of safety and security in line with the regulatory intent. Where quality standards are intended to be regulated for deceptive trade practices, the standards should contain performance requirements and should not restrict input materials or process routes. Regulators should also ensure that the standards used regulations which are adopted or closely harmonized with international standards. In these cases, where essential requirements are directly incorporated in the technical requlations, these should be drawn by empanelled experts.

Technical regulations should select conformity assessment procedures that are least trade restrictive and burdensome to cover the breadth and magnitude of the safety, security or deception risks and take into account the available conformity assessment bodies/laboratory infrastructure. Presently regulations that rely on conformity assessment as the means for compliance have with a few exceptions, chosen product certification routes operated by BIS or by the regulators themselves, which impose a high cost of compliance. As several other conformity assessment routes are available such as self-declaration of conformity design certification, batch certification, third-party inspection, sample testing, capability approvals, these should be examined for best suitability before selection. Preference may be accorded to self-declaration of conformity but these should be combined with adequate prior and post-market testing, and backed by stringent penal provisions in case of wilful default. Technical regulations should be finalized only after wide stakeholder consultation and fulfilment of WTO notification obligations.

Further SPS measures should be based on ISGRs. If ISGRs are not sufficient to meet the desired level of protection, the SPS measures should be backed by scientific principles and sufficient scientific evidence and be based on risk assessment where scientific evidence is not sufficient measures could be provisionally adopted on the basis of available justification.

Goal 5: Create an overarching regulatory instrument and oversight mechanism for technical regulations

Presently technical regulations are notified under different Acts, and under the BIS Act where no sector specific Act is existing. These Acts have been drafted to serve specific purposes and are limited in their scope. It is essential to enact a new enabling legislation for notifying standards, technical regulations, and conformity assessment procedures in accordance with global good regulatory practices with suitable surveillance and enforcement provisions. This legislation should apply to all those sectors that are presently not covered by sector-specific regulations and should have needed. It should also include provisions for regulatory impact assessment and periodical reviews and sunset clauses.

A suitable regulatory instrument is also required to protect consumers against unsupported claims of conformity by suppliers or conformity assessment bodies that are not accredited and therefore not accountable. The market surveillance mechanism should have a provision for cross-border exchange of information.

Goal 6:Create an effective market surveillance mechanism

Presently market surveillance activities and other enforcement measures are handled by state Government agencies and customs officials at the ports who are not best equipped in terms of technical understanding, resources and empowerment. As the requirements for post-market surveillance and testing including cyber intelligence are expected to increase in future, it is necessary to establish a professional agency for carrying out or coordinating the market surveillance programme (recommended in Conformity Assessment Goal 1) and port control operations. Market surveillance should invariably include testing of products drawn from the market and in cases of wilful deceptive practices, and statutory actions to prevent further supplies.

Goal 7:Strengthening response mechanisms to overseas technical regulations and SPS measures

New and revised technical regulations and SPS measures are being regularly notified by countries classified as TBT and SPS notifications. Many of these have a direct or indirect impact on the supply of Indian goods and services. An urgent need exists to develop a dynamic and responsive mechanism that would trigger anticipatory as well as post notification responses. As a first step, a strengthening of institutional arrangements is required along with a creation of a digital platform to analyse all new notifications, bring them to the knowledge of the corresponding regulators and impacted suppliers (including potential suppliers), collate their concerns and issue timely national responses to the notification the role of and active participation from the subject matter ministries as well as industry bodies is critical in strengthening the national response mechanism. A strengthened response system will also help regulators understand best regulatory practices and the regulatory gap in the country.

Post notification responses include understanding the impact in terms of resources, technology readiness, and quality infrastructure required to meet the compliance requirements. Depending on the magnitude of impact, subject matter ministries, relevant departments, and bodies under them should prepare an impact document indicating the assistance at the government level that should be provided, especially to the MSMEs.

AWARENESS , COUNSELLING, TRAINING AND EDUCATION (ACT & E)

Vision: Creating a quality mindset nationwide.

Mission: Make every citizen, organization, and institution understand, and value the benefits that standardization and related activities bring to them.

Goal 1: Enhancing awareness among stakeholders

Standards, conformity assessment, and technical regulations when applied are routinely encountered by all citizens, organizations, and institutions in their daily lives and operations. However, lack of awareness on their relevance leads to limited participation in standards setting and conformity assessment processes, and inability to derive their full potential and benefits as suppliers, service providers, consumers, policymakers and regulators. An objective of this strategy for standardization is to widely transmit the message to all stakeholders so that not only do they realize the opportunities but also become responsible role players through active participation. It is also necessary to create awareness of specific sectoral conformity assessment requirements imposed through national and global schemes that have an impact on domestic supply and export.

The target audience for building awareness should include officials responsible for policy and technical regulations, the enforcement agencies, public procurement agencies both under central and state governments, officials responsible for port controls, trade and industry, and consumer organizations. Consumer awareness should include where and how complaints, disputes, and appeals can be raised.

Goal 2: Counselling and training regarding standards; conformity assessment; technical regulations, and SPS aspects

The qualified ecosystem in India comprises of a sizable body of knowledge in the form of standards, multiple conformity assessment programmes, and national schemes. There is also an increasing number of technical regulations and SPS measures put in place by the importing countries, besides myriad requirements of a buyer in foreign markets that affect trade. Producers of goods and services, and their importers and exporters may require constant counselling in this regard. Such counselling will be expected to help the exporters to understand the differences between buyer's requirement and mandatory requirements (in the form of directives/technical regulations in the importing country). Trade organizations like CII, FICCI ASSOCHAM, FIEO, etc., will need to calibrate their counselling services to the interested stakeholders aligned with these needs. The trade portal (managed by FIEO) and India Standards Portal (managed by CII) also need to enhance their scope as hubs of information in this regard.

Given the complexities and the technical nature of the regulatory eco-system in international trade, it is essential that government officials, regulators, enforcement agencies and monitoring agencies are adequately skilled and trained from time to time so as to handle the responsibilities expected of them. The training modules should be the joint responsibility of the regulators, BIS, the accreditation boards, trade promotion boards/councils and the apex industry associations with assistance from the State government academies of administration. The National Institute of Training for Standardization of BIS should be made responsible for creating training and awareness modules for enhancing awareness up to District levels and industrial clusters.

Public procurement officials need to be sensitized on according precedence to national standards and conformity assessment schemes.

A major role in disseminating standards-based information directly to a user is played by professional trainers and consultants. This task is best done in the non-governmental sector, in an environment where the training/counselling sector is free to grow as per market demand. A system for identifying and certification of competent trainers and consultants and creation of a corresponding accreditation system for training/consulting organizations and certification system for trainers/consultants in different sectors exist under the aegis of Quality Council of India, which needs to be enlarged to cover all streams of standards and conformity assessment schemes.

Goal 3: Creating modules for courses on quality related subjects in educational institutions at various levels

One of the most expedient tasks for building the quality mindset into the potential workforce involves building suitable curriculum and its inclusion in the various stages of formal education. This would entail the development of specific modules related to standardization, quality practices, and statutory provisions protecting consumer rights, etc., in educational programmes from primary to higher education levels with commensurate depth. The education as a minimum should include basic standards that impact society at large such as metrology based on SI units, concepts of interoperability, standards addressing human health and safety, and environment.

As standards are extensively used in engineering and industrial applications, it is essential to integrate the related professional courses with the vast contemporary technical knowledge that standards provide. This will enable the students to be "industry- ready" on graduation. A strong connect between standards bodies and the technical/professional institutes is essential for which policy guidelines need to be issued by the Ministry of Human Resource Development and made part of the course accreditation criteria.

LIST OF ABBREVIATIONS

ASSOCHAM	The Associated Chambers of Commerce and Industry of India
ASI	Archaeological Survey of India
BIPM	International Bureau of Weights and Measures
BIS	Bureau of Indian Standards
CAB	Conformity Assessment Bodies
CASCO	Committee on Conformity Assessment
CII	Confederation of Indian Industries

CTIL	Centre for Trade and Investment Law
CWTOS	Centre for WTO Studies
EIC	Export Inspection Council
FICCI	Federation of Indian Chambers of Commerce and Industry
FIEO	Federation of Indian Export Organisations
FSSAI	Food Safety and Standards Authority of India
IAF	International Accreditation Forum
IBEF	Indian Brand Equity Foundation
IEC	International Electrotechnical Commission
INSS	Indian National Strategy for Standardization
IPCC	International Plant Protection Convention
ISGRs	International Standards, Guidelines and Recommendations
ISO	International Organization for Standardization
ISRO	Indian Space Research Organisation
ITU	International Telecommunication Union
ILAC	International Laboratory Accreditation Cooperation
MSMEs	Micro, Small, and Medium Enterprises
NABCB	National Accreditation Board for Certification Bodies
NABL	National Accreditation Board for Testing and Calibration Laboratories
PSS	Private Sustainability Standards
SAARC	South Asian Association for Regional Cooperation
SARSO	South Asian Regional Standards Organization
SDOs	Standards Development Organizations
SDGs	Sustainable Development Goals
SI	International System of Units
SPS	Agreement on Sanitary and Phytosanitary Measures
TBT	Agreement on Technical Barriers to Trade
TPD	Trade Policy Division, Department of Commerce
VSS	Voluntary Sustainability Standards

WTO　　　　　World Trade Organization

ACKNOWLEDGEMENT

This Strategy document is prepared by Joint Secretary Shri Sudhanshu Pandey under overall supervision of Officer on Special Duty Dr. Anup Wadhawan of Trade Policy Division of Department of Commerce after extensive consultation with various stakeholders and seeking comments from general public on the draft document. The Division acknowledges the contribution of Bureau of Indian Standards (BIS), subject experts, industry associations, export promotion boards/agencies, ministries/ departments/regulators, state governments and general public for their comments/ suggestions/feedbacks.

In finalizing the document, special acknowledgement is due to Shri Anupam Kaul, Head-Quality, Metrology and Standards (QMS), CII, for initiating the foundational text of the document. Acknowledgement is also due to the members of the expert group who perused the numerous comments from the stakeholders and helped the strategy reach its final shape, viz. Shri Anil Jauhri, CEO, NABCB, Prof. Murali Kallummal (from CWTOS), Dr. James Nedumpara and Aditya Laddha (from CTIL), Shri T. S. Vishwanath (Principal Adviser, APJ-SLG, Law Offices), Shri Pramod Siwach (EIC), Ms. Sangeeta Saxena (TPD, Services) and Shri Pranav Kumar (CII).

Contribution of Shipra Abraham and Nisha Kathait from the departmental social media cell for widely publicizing the draft document during its public comment stage and that of Shoumi Dasgupta and Preeti Handa (from IBEF) for designing the booklet is also acknowledged. Jhanvi Tripathi (CII) needs special mention for keeping track of numerous versions of the document and being a strict editor of words.

The Division would like to express deep gratitude to the Commerce Secretary, Ms. Rita Teaotia, for her constant guidance on the broad as well as fine contours of the strategy. Without her support it would not have been possible to ensure coordination and cooperation from various stakeholders.

Above all, the commanding vision of Honorable Commerce & Industries Minister Shri Suresh Prabhu was the constant source of inspiration for the policy articulation and his indefatigable spirit guided every one of us in preparing this strategy document.

附录 2　《印度标准局法案 2016》英文原文

THE BUREAU OF INDIAN STANDARDS ACT, 2016

——

ARRANGEMENT OF SECTIONS

——

CHAPTER I

PRELIMINARY

12. Conformity Assessment scheme.

13. Grant of licence or certificate of conformity.

14. Certification of Standard Mark of jewellers and sellers of certain specified goods or articles.

15. Prohibition to import, sell, exhibit, etc.

16. Central Government to direct compulsory use of Standard Mark.

17. Prohibition to manufacture, sell, etc., certain goods without Standard Mark.

18. Obligations of licence holder, seller, etc.

CHAPTER IV

FINANCE, ACCOUNTS AND AUDIT

19. Financial Management of Bureau of Indian Standards.

20. Fund of Bureau.

21. Borrowing powers of Bureau.

22. Budget.

23. Annual report.

24. Accounts and audit.

CHAPTER V

MISCELLANEOUS

25. Power of Central Government to issue directions.

26. Restriction on use of name of Bureau and Indian Standard.

27. Appointment and powers of certification officers.

28. Power to search and seizure.

29. Penalty for contravention.

30. Offences by companies.

31. Compensation for non-conforming goods.

32. Cognizance of offence by courts.

33. Compounding of offence.

34. Appeal.

THE BUREAU OF INDIAN STANDARDS ACT, 2016

ACT NO. 11 OF 2016

[21st March, 2016]

An Act to provide for the establishment of a national standards body for the harmonious development of the activities of standardisation, conformity assessment and quality assurance of goods, articles, processes, systems and services and for matters connected therewith or incidental thereto.

Be it enacted by Parliament in the Sixty-seventh Year of the Republic of India as follows:

CHAPTER I

PRELIMINARY

1. Short title, extent and commencement

(1) This Act may be called the Bureau of Indian Standards Act, 2016.

(2) It extends to the whole of India.

(3) It shall come into force on such 1 date as the Central Government may, by notification in the Official Gazette, appoint.

2. Definitions

In this Act, unless the context otherwise requires.

(1) "article" means any substance, artificial or natural, or partly artificial or partly natural, whether raw or partly or wholly processed or manufactured or

handmade within India or imported into India;

(2) "assaying and hallmarking centre" means a testing and marking centre recognised by the Bureau to determine the purity of precious metal articles and to apply hallmark on the precious metal articles in a manner as may be determined by regulations;

(3) "Bureau" means the Bureau of Indian Standards established under section 3;

(4) "certification officer" means a certification officer appointed under sub-section (1) of section 27;

(5) "certified body" means a holder of certificate of conformity or licence under sub-section (2) of section 13 in relation to any goods, article, process, system or service which conforms to a standard;

(6) "certified jeweller" means a jeweller who has been granted a certificate by the Bureau to get manufactured for sale or to sell any precious metal article after getting the same hallmarked in a manner as may be determined by regulations;

(7) "conformity assessment" means demonstration that requirements as may be specified relating to an article, process, system, service, person or body are fulfilled;

(8) "conformity assessment scheme" means a scheme relating to such goods, article, process, system or service as may be notified by the Bureau under section 12;

(9) "consumer" means a person as defined in the Consumer Protection Act, 1986 (68 of 1986);

(10) "covering" includes any stopper, cask, bottle, vessel, box, crate, cover, capsule, case, frame, wrapper, bag, sack, pouch or other container;

(11) "Director General" means the Director General appointed under sub-section (1) of section 7;

(12) "Executive Committee" means the Executive Committee constituted under sub-section (1) of section 4;

(13) "fund" means the fund constituted under section 20;

(14) "goods" includes all kinds of movable properties under the Sale of Goods Act, 1930 (3 of 1930), other than actionable claims, money, stocks and shares;

(15) "Governing Council" means a Governing Council constituted under sub-section (3) of section 3;

(16) "Hallmark" means in relation to precious metal article, the Standard Mark, which indicates the proportionate content of precious metal in that article as per the relevant Indian Standard;

(17) "Indian Standard" means the standard including any tentative or provisional standard established and published by the Bureau, in relation to any goods, article, process, system or service, indicative of the quality and specification of such goods, article, process, system or service and includes:

(a) any standard adopted by the Bureau under sub-section (2) of section 10;

(b) any standard established and published, or recognised, by the Bureau of Indian Standards established under the Bureau of Indian Standard Act, 1986 (63 of 1986), which was in force immediately before the commencement of this Act;

(18) "Indian Standards Institution" means the Indian Standards Institution registered under the Societies Registration Act, 1860 (21 of 1860);

(19) "jeweller" means a person engaged in the business to get manufactured precious metal article for sale or to sell precious metal articles;

(20) "licence" means a licence granted under section 13 to use a specified Standard Mark in relation to any goods, article, process, system or service, which conforms to a standard;

(21) "manufacturer" means a person responsible for designing and manufacturing any goods or article;

(22) "mark" includes a device, brand, heading, label, ticket, pictorial representation, name, signature, word, letter or numeral or any combination thereof;

(23) "member" means a member of the Governing Council, Executive Committee or any of the Advisory Committee;

(24) "notification" means a notification published in the Official Gazette and the expression "notify" or "notified" shall be construed accordingly;

(25) "person" means a manufacturer, an importer, a distributor, retailer, seller

or lessor of goods or article or provider of service or any other person who uses or applies his name or trade mark or any other distinctive mark on to goods or article or while providing a service, for any consideration or gives goods or article or provides service as prize or gift for commercial purposes including their representative and any person who is engaged in such activities, where the manufacturer, importer, distributor, retailer, seller, lessor or provider of service cannot be identified;

(26) "precious metal" means gold, silver, platinum and palladium;

(27) "precious metal article" means any article made entirely or in part from precious metals or their alloys;

(28) "prescribed" means prescribed by rules made under this Act;

(29) "process" means a set of inter-related or interacting activities, which transforms inputs into outputs;

(30) "recognised testing and marking centre" means a testing and marking centre recognised by the Bureau under sub-section (5) of section 14;

(31) "recognised testing laboratory" means a testing laboratory recognised by the Bureau under sub-section (4) of section 13;

(32) "registering authority" means any authority competent under any law for the time being in force to register any company, firm or other body of persons, or any trade mark or design, or to grant a patent;

(33) "regulations" means regulations made by the Bureau under this Act;

(34) "sale" means to sell, distribute, hire, lease or exchange of goods, article, process, system or service for any consideration or for commercial purposes;

(35) "seller" means a person who is engaged in the sale of any goods, article, process, system or service;

(36) "service" means the result generated by activities at the interface between an organisation and a customer and by organisation's internal activities, to meet customer requirements;

(37) "specification" means a description of goods, article, process, system or service as far as practicable by reference to its nature, quality, strength, purity,

composition, quantity, dimensions, weight, grade, durability, origin, age, material, mode of manufacture or processing, consistency and reliability of service delivery or other characteristics to distinguish it from any other goods, article, process, system or service;

(38) "specified" means specified by the regulations;

(39) "standards" means documented agreements containing technical specifications or other precise criteria to be used consistently as rules, guidelines, or definitions of characteristics, to ensure that goods, articles, processes, systems and services are fit for their purpose;

(40) "Standard Mark" means the mark specified by the Bureau, and includes Hallmark, to represent conformity of goods, article, process, system or service to a particular Indian Standard or conformity to a standard, the mark of which has been established, adopted or recognised by the Bureau and is marked on the article or goods as a Standard Mark or on its covering or label attached to such goods or article so marked;

(41) "system" means a set of inter-related or interacting elements;

(42) "testing laboratory" means a body set up for the purpose of testing of goods or article against a set of requirements and report its findings;

(43) "trade mark" means a mark used or proposed to be used in relation to goods or article or process or system or service for the purpose of indicating, or so as to indicate, a connection in the course of trade of goods, article, process, system or service, as the case may be, and some person having the right, either as proprietor or as registered user, to use the mark, whether with or without any indication of the identity of that person.

CHAPTER II

BUREAU OF INDIAN STANDARDS

3. Establishment of Bureau and Constitution of Governing Council

(1) With effect from such date as the Central Government may, by notification in the Official Gazette, appoint in this behalf, there shall be established a national body

for the purposes of this Act, a Bureau, to be called the Bureau of Indian Standards.

(2) The Bureau shall be a body corporate by the name aforesaid, having perpetual succession and a common seal, with power, subject to the provisions of this Act, to acquire, hold and dispose of property, both movable and immovable, and to contract and shall by the said name sue and be sued.

(3) The members of the Governing Council shall constitute the Bureau and general superintendence, direction and management of the affairs of the Bureau shall vest in the Governing Council, which shall consist of the following members, namely:

(a) the Minister in-charge of the Ministry or Department of the Central Government having administrative control of the Bureau who shall be ex officio President of the Bureau;

(b) the Minister of State or a Deputy Minister, if any, in the Ministry or Department of the Central Government having administrative control of the Bureau who shall be ex officio Vice-President of the Bureau, and where there is no such Minister of State or Deputy Minister, such person as may be nominated by the Central Government to be the Vice-President of the Bureau;

(c) the Secretary to the Government of India of the Ministry or Department of the Central Government having administrative control of the Bureau, ex officio;

(d) the Director General of the Bureau, ex officio;

(e) such number of other persons to represent the Government, industry, scientific and research institutions, consumers and other interests, as may be prescribed, to be appointed by the Central Government.

(4) The term of office of the members referred to in clause (e) of sub-section (3) and the manner of filling vacancies among, and the procedure to be followed in the discharge of their functions by the members, shall be such as may be prescribed:

Provided that a member, other than an ex officio member of the Bureau of Indian Standards constituted under the Bureau of Indian Standards Act, 1986 (63 of 1986), shall, after the commencement of this Act, continue to hold such office as member till the completion of his term.

(5) The Governing Council may associate with itself, in such manner and for such purposes as may be prescribed, any person whose assistance or advice it may desire in complying with any of the provisions of this Act and a person so associated shall have the right to take part in the discussions of the Governing Council relevant to the purposes for which he has been associated but shall not have the right to vote.

(6) The Governing Council may, by general or special order in writing, delegate to any member, the Director General or any other person subject to such conditions, if any, as may be specified in the order, such of its powers and functions under this Act except the powers under section 37 as it may deem necessary.

4. Executive Committee of Bureau

(1) The Governing Council may, with the prior approval of the Central Government, by notification in the Official Gazette, constitute an Executive Committee which shall consist of the following members, namely:

(a) Director General of the Bureau, who shall be its ex officio Chairman;

(b) such number of members, as may be prescribed.

(2) The Executive Committee constituted under sub-section (1) shall perform, exercise and discharge such functions, powers and duties of the Bureau, as may be delegated to it by the Governing Council.

5. Advisory Committees of Bureau

(1) Subject to any regulations made in this behalf, the Governing Council may, from time to time and as and when it is considered necessary, constitute the following Advisory Committees for the efficient discharge of the functions of the Bureau, namely:

(a) Finance Advisory Committee;

(b) Conformity Assessment Advisory Committee;

(c) Standards Advisory Committee;

(d) Testing and Calibration Advisory Committee;

(e) such number of other committees as may be specified by regulations.

(2) Each Advisory Committee shall consist of a Chairman and such other members as may be specified by regulations.

6. Vacancies, etc. , not to invalidate act or proceedings —No act or proceedings of the Governing Council, under section 3 shall be invalid merely by reason of—

(a) any vacancy in, or any defect in the constitution of the Governing Council; or

(b) any defect in the appointment of a person acting as a member of the Governing Council; or

(c) any irregularity in the procedure of the Governing Council not affecting the merits of the case.

7. Director General

(1) The Central Government shall appoint a Director General of the Bureau.

(2) The terms and conditions of service of the Director General of the Bureau shall be such manners as may be prescribed.

(3) Subject to the general superintendence and control of the Governing Council, the Director General of the Bureau shall be the Chief Executive Authority of the Bureau.

(4) The Director General of the Bureau shall exercise and discharge such of the powers and duties of the Bureau as may be specified by regulations.

(5) The Director General may, by general or special order in writing, delegate to any officer of the Bureau subject to such conditions, if any, as may be specified in the order, such of his powers and functions as are assigned to him under the regulations or are delegated to him by the Governing Council, as he may deem necessary.

8. Officers and employees of Bureau

(1) The Bureau may appoint such other officers and employees as it considers necessary for the efficient discharge of its functions under this Act.

(2) The terms and conditions of service of officers and employees of the Bureau appointed under sub-section (1) shall be such as may be specified by regulations.

9. Powers and functions of Bureau

(1) The powers and duties as may be assigned to the Bureau under this Act shall be exercised and performed by the Governing Council and, in particular, such powers may include the power to—

(a) establish branches, offices or agencies in India or outside;

(b) recognise, on reciprocal basis or otherwise, with the prior approval of the Central Government, the mark of any international body or institution, on such terms and conditions as may be mutually agreed upon by the Bureau in relation to any goods, article, process, system or service at par with the Standard Mark for such goods, article, process, system or service;

(c) seek recognition of the Bureau and of the Indian Standards outside India on such terms and conditions as may be mutually agreed upon by the Bureau with any corresponding institution or organisation in any country or with any international organisation;

(d) enter into and search places, premises or vehicles, and inspect and seize goods or articles and documents to enforce the provisions of this Act;

(e) provide services to manufacturers and consumers of goods or articles or processes for compliances of standards on such terms and conditions as may be mutually agreed upon;

(f) provide training services in relation to quality management, standards, conformity assessment, laboratory testing and calibration, and any other related areas;

(g) publish Indian Standards and sell such publications and publications of international bodies;

(h) authorise agencies in India or outside India for carrying out any or all activities of the Bureau and such other purposes as may be necessary on such terms and conditions as it deems fit;

(i) obtain membership in regional, international and foreign bodies having objects similar to that of the Bureau and participate in international standards setting process;

(j) undertake testing of samples for purposes other than for conformity assessment;

(k) undertake activities relating to legal metrology.

(2) The Bureau shall take all necessary steps for promotion, monitoring and management of the quality of goods, articles, processes, systems and services, as may

be necessary, to protect the interests of consumers and various other stakeholders which may include the following namely:

(a) carrying out market surveillance or survey of any goods, article, process, system or service to monitor their quality and publish findings of such surveillance or surveys;

(b) promotion of quality in connection with any goods, article, process, system or service by creating awareness among the consumers and the industry and educate them about quality and standards in connection with any goods, article, process, system and service;

(c) promotion of safety in connection with any goods, article, process, system or service;

(d) identification of any goods, articles, process, system or service for which there is a need to establish a new Indian Standard, or to revise an existing Indian Standard;

(e) promoting the use of Indian Standards;

(f) recognising or accrediting any institution in India or outside which is engaged in conformity certification and inspection of any goods, article, process, system or service or of testing laboratories;

(g) coordination and promotion of activities of any association of manufacturers or consumers or any other body in relation to improvement in the quality or in the implementation of any quality assurance activities in relation to any goods, article, process, system or service;

(h) such other functions as may be necessary for promotion, monitoring and management of the quality of goods, articles, processes, systems and services and to protect the interests of consumers and other stakeholders.

(3) The Bureau shall perform its functions under this section through the Governing Council in accordance with the direction and subject to such rules as may be made by the Central Government.

CHAPTER III

INDIAN STANDARDS, CERTIFICATION AND LICENCE

10. Indian Standards

(1) The standards established by the Bureau shall be the Indian Standards.

(2) The Bureau may—

(a) establish, publish, review and promote the Indian Standard, in relation to any goods, article, process, system or service in such manner as may be prescribed;

(b) adopt as Indian Standard, any standard, established by any other Institution in India or elsewhere, in relation to any goods, article, process, system or service in such manner as may be prescribed;

(c) recognise or accredit any institution in India or outside which is engaged in standardisation;

(d) undertake, support and promote such research as may be necessary for formulation of Indian Standards.

(3) The Bureau, for the purpose of this section, shall constitute, as and when considered necessary, such number of technical committees of experts for the formulation of standards in respect of goods, articles, processes, systems or services, as may be necessary.

(4) The Indian Standard shall be notified and remain valid till withdrawn by the Bureau.

(5) Without understanding anything contained in any other law, the copyright in an Indian Standard or any other publication of the Bureau shall vest in the Bureau.

11. Prohibition to publish, reproduce or record without authorisation by Bureau

(1) No individual shall, without the authorisation of the Bureau, in any manner or form, publish, reproduce or record any Indian Standard or part thereof, or any other publication of the Bureau.

(2) No person shall issue a document that creates, or may create the impression that it is or contains an Indian Standard, as contemplated in this Act:

Provided that nothing in this sub-section shall prevent any individual from making a copy of Indian Standard for his personal use.

12. Conformity Assessment scheme

(1) The Bureau may notify a specific or different conformity assessment scheme for any goods, article, process, system or service or for a group of goods, articles, processes, systems or services, as the case may be, with respect to any Indian Standard or any other standard in a manner as may be specified by regulations.

(2) The Bureau may establish a Standard Mark in relation to each of its conformity assessment schemes, which shall be of such design and contain such particulars as may be specified by regulations to represent a particular standard.

13. Grant of licence or certificate of conformity

(1) A person may apply for grant of licence or certificate of conformity, as the case may be, if the goods, article, process, system or service conforms to an Indian Standard.

(2) Where any goods, article, process, system or service conforms to a standard, the Director General may, by an order, grant—

(a) a certificate of conformity in a manner as may be specified by regulations; or

(b) a licence to use or apply a Standard Mark in a manner as may be specified by regulations, subject to such conditions and on payment of such fees, including late fee or fine, before or during the operation of the certificate of conformity or licence, and as determined by regulations.

(3) While granting a certificate of conformity or licence to use a Standard Mark, the Bureau may, by order, specify the marking and labelling requirements that shall necessarily be affixed as may be specified from time to time.

(4) The Bureau may establish, maintain or recognise testing laboratories for the purposes of conformity assessment and quality assurance and for such other purposes as may be required for carrying out its functions.

14. Certification of Standard Mark of jewellers and sellers of certain specified goods or articles

(1) The Central Government, after consulting the Bureau, may notify precious

metal articles or other goods or articles as it may consider necessary, to be marked with a Hallmark or Standard Mark, as the case may be, in a manner as specified in sub-section (2).

(2) The goods or articles notified in sub-section (1) may be sold through retail outlets certified by the Bureau after such goods or articles have been assessed for conformity to the relevant standard by testing and marking centre, recognised by the Bureau and marked with Hallmark or Standard Mark, as the case may be, as specified by regulations.

(3) The Central Government may, after consulting the Bureau, by an order published in the Official Gazette, make it compulsory for the sellers of goods or article notified under sub-section (1) to be sold only through certified sales outlets fulfilling such conditions as may be determined by regulations.

(4) The Bureau may, by an order, grant, renew, suspend or cancel certification of Standard Mark or Hallmark of a jeweller or any other seller for sale of goods or articles notified under sub-section (1) in such manner as may be determined by regulations.

(5) The Bureau may establish, maintain and recognise testing and marking centres, including assaying and hallmarking centres, for conformity assessment and application of Standard Mark, including Hallmark, on goods or articles notified under sub-section (1), in a manner as may be specified by regulations.

(6) No testing and marking centre or assaying and hallmarking centre, other than the recognised by the Bureau, shall with respect to goods or articles notified under sub-section (1), use, affix, emboss, engrave, print or apply in any manner the Standard Mark, including the Hallmark, or colourable imitation thereof, on any goods or article; and make any claim in relation to the use and application of a Standard Mark, including the Hallmark, through advertisements, sales promotion leaflets, price lists or the like.

(7) Every recognised testing and marking centre, including assaying and hallmarking centre, shall use or apply Standard Mark on goods or articles notified

under sub-section (1), including Hallmark on precious metal articles, after accurately determining the conformity of the same in a manner as may be specified.

(8) No recognised testing and marking centre, including assaying and hallmarking centre, shall, notwithstanding that it has been recognised under sub-section (5), use or apply in relation to any goods or article notified under sub-section (1) a Standard Mark, including Hallmark, or any colourable imitation thereof, unless such goods or article conforms to the relevant standard.

15. Prohibition to import, sell, exhibit, etc

(1) No person shall import, distribute, sell, store or exhibit for sale, any goods or article under sub-section (1) of section 14, except under certification from the Bureau.

(2) No person, other than that certified by the Bureau, shall sell or display or offer to sell goods or articles that are notified under sub-section (3) of section 14 and marked with the Standard Mark, including Hallmark and claim in relation to the Standard Mark, including Hallmark, through advertisements, sales promotion leaflets, price lists or the like.

(3) No certified jeweller or seller shall sell or display or offer to sell any notified goods or articles, notwithstanding that he has been granted certification, with the Standard Mark, including Hallmark, or any colourable imitation thereof, unless such goods or article is marked with a Standard Mark or Hallmark, in a manner as may be specified by regulations, and unless such goods or article conforms to the relevant standard.

16. Central Government to direct compulsory use of Standard Mark

(1) If the Central Government is of the opinion that it is necessary or expedient so to do in the public interest or for the protection of human, animal or plant health, safety of the environment, or prevention of unfair trade practices, or national security, it may, after consulting the Bureau, by an order published in the Official Gazette, notify—

(a) goods or article of any scheduled industry, process, system or service; or

(b) essential requirements to which such goods, article, process, system or service, which shall conform to a standard and direct the use of the Standard Mark under a licence or certificate of conformity as compulsory on such goods, article, process, system or service.

Explanation. —For the purpose of this sub-section—

(i) the expression "scheduled industry" shall have the meaning assigned to it in the Industries (Development and Regulation) Act, 1951 (65 of 1951);

(ii) it is hereby clarified that essential requirements are requirements, expressed in terms of the parameters to be achieved or requirements of standard in technical terms that effectively ensure that any goods, article, process, system or service meet the objective of health, safety and environment.

(2) The Central Government may, by an order authorise Bureau or any other agency having necessary accreditation or recognition and valid approval to certify and enforce conformity to the relevant standard or prescribed essential requirements under sub-section (1).

17. Prohibition to manufacture, sell, etc. , certain goods without Standard Mark

(1) No person shall manufacture, import, distribute, sell, hire, lease, store or exhibit for sale any such goods, article, process, system or service under sub-section (1) of section 16—

(a) without a Standard Mark, except under a valid licence; or

(b) notwithstanding that he has been granted a license, apply a Standard Mark, unless such goods, article, process, system or service conforms to the relevant standard or prescribed essential requirements.

(2) No person shall make a public claim, through advertisements, sales promotion leaflets, price lists or the like, that his goods, article, process, system or service conforms to an Indian standard or make such a declaration on the goods or article, without having a valid certificate of conformity or licence from the Bureau or any other authority approved by the Central Government under sub-section (2) of section 16.

(3) No person shall use or apply or purport to use or apply in any manner, in the manufacture, distribution, sale, hire, lease or exhibit or offer for sale of any goods, article, process, system or service, or in the title of any patent or in any trade mark or design, a Standard Mark or any colourable imitation thereof, except under a valid licence from the Bureau.

18. Obligations of licence holder, seller, etc

(1) The licence holder shall, at all times, remain responsible for conformance of the goods, articles, processes, systems or services carrying the Standard Mark.

(2) It shall be the responsibility of the distributor or the seller, as the case may be, to ensure that goods, articles, processes, systems or services carrying the Standard Mark are purchased from certified body or licence holder.

(3) It shall be the responsibility of the seller before the goods or article is sold or offered to be sold or exhibited or offered for sale to ensure that—

(a) goods, articles, processes, systems or services carrying the Standard Mark bear the requisite labels and marking details, as specified by the Bureau from time to time;

(b) the marking and labelling requirements on the product or covering is displayed in a manner that has been specified by the Bureau.

(4) Every certified body or licence holder shall supply to the Bureau with such information and with such samples of any material or substance used in relation to any goods, article, process, system or service, as the case may be, as the Bureau may require for monitoring its quality and for the recovery of the fee as may be prescribed in the certificate of conformity or the licence.

(5)

(a) The Bureau may make such inspection and take such samples of any material or substance as may be necessary to see whether any goods, article, process, system or service, in relation to which a Standard Mark has been used, conforms to the requirements of the relevant standard or whether the Standard Mark has been properly used in relation to any goods, article, process, system or service with or without a licence.

(b) The Bureau may publicise the results of its findings and the directions given in pursuance thereof.

(6) If the Bureau is satisfied under the provisions of sub-sections (4) and (5) that the goods, articles, processes, systems or services in relation to which a Standard Mark has been used do not conform to the requirements of the relevant standard, the Bureau may direct the certified body or licence holder or his representative to stop the supply and sale of non-conforming goods or articles and recall the non-conforming goods or articles that have already been supplied or offered for sale and bear such mark from the market or any such place from where they are likely to be offered for sale or prohibit to provide the service.

(7) Where a certified body or licence holder or his representative has sold goods, articles, processes, systems or services, which bear a Standard Mark or any colourable imitation thereof, which do not conform to the relevant standard, the Bureau shall direct the certified body or licence holder or his representative to—

(a) repair or replace or reprocess the standard marked goods, article, process, system or service in a manner as may be specified; or

(b) pay compensation to the consumer as may be prescribed by the Bureau; or

(c) be liable for the injury caused by non-conforming goods or article, which bears a Standard Mark, as per the provisions of section 31.

CHAPTER IV

FINANCE, ACCOUNTS AND AUDIT

19. Financial Management of Bureau of Indian Standards

The Central Government may, after due appropriation made by Parliament by law in this behalf, make to the Bureau grants and loans of such sums of money as the Government may consider necessary.

20. Fund of Bureau

(1) There shall be constituted a fund to be called the Bureau of Indian Standards fund and there shall be credited there to—

(a) any grants and loans made to the Bureau by the Central Government;

(b) all fees and charges received by the Bureau under this Act;

(c) all fines received by the Bureau;

(d) all sums received by the Bureau from such other sources as may be decided upon by the Central Government.

(2) The fund shall be applied for meeting—

(a) the salary, allowances and other remuneration of the members, Director General, officers and other employees of the Bureau;

(b) expenses of the Bureau in the discharge of its functions under the Act;

(c) expenses on objects and for purposes authorised by this Act:

Provided that the fines received in clause (c) of sub-section (1) shall be used for consumer awareness, consumer protection and promotion of quality of goods, articles, processes, systems or services in the country.

21. Borrowing powers of Bureau

(1) The Bureau may, with the consent of the Central Government or in accordance with the terms of any general or special authority given to it by the Central Government, borrow money from any source as it may deem fit for discharging all or any of its functions under this Act.

(2) The Central Government may guarantee in such manner as it thinks fit, the repayment of the principal and the payment of interest thereon with respect to the loans borrowed by Bureau undersub-section (1).

22. Budget

The Bureau shall prepare, in such form and at such time in each financial year as may be prescribed, its budget for the next financial year, showing the estimated receipts and expenditure of the Bureau and forward the same to the Central Government.

23. Annual report

(1) The Bureau shall prepare, in such form and at such time in each financial year as may be prescribed, its annual report, giving a full account of its activities during

the previous financial year, and submit a copy thereof to the Central Government.

(2) The Central Government shall cause the annual report to be laid, as soon as may be after it is received, before each House of Parliament.

24. Accounts and audit

(1) The Bureau shall maintain proper accounts and other relevant records and prepare an annual statement of accounts, in such form as may be prescribed by the Central Government in consultation with the Comptroller and Auditor-General of India.

(2) The accounts of the Bureau shall be audited by the Comptroller and Auditor-General of India at such intervals as may be specified by him and any expenditure incurred in connection with such audit shall be payable by the Bureau to the Comptroller and Auditor-General of India.

(3) The Comptroller and Auditor-General of India and any person appointed by him in connection with the audit of the accounts of the Bureau shall have the same rights and privileges and the authority in connection with such audit as the Comptroller and Auditor-General of India generally has in connection with the audit of Government accounts and, in particular, shall have the right to demand the production of books, accounts, connected vouchers and other documents and papers and to inspect any office of the Bureau.

(4) The accounts of the Bureau as certified by the Comptroller and Auditor-General of India or any other person appointed by him in this behalf together with the audit report thereon shall be forwarded annually to the Central Government and that Government shall cause the same to be laid before each House of Parliament.

CHAPTER V

MISCELLANEOUS

25. Power of Central Government to issue directions

(1) Without prejudice to the foregoing provisions of this Act, the Bureau shall, in the exercise of its powers or the performance of its functions under this Act, be bound by such directions on questions of policy as the Central Government may give in

writing to it from time to time:

Provided that the Bureau shall, as far as practicable, be given an opportunity to express its views before any direction is given under this sub-section.

(2) The decision of the Central Government whether a question is one of policy or not shall be final.

(3) The Central Government may take such other action as may be necessary for the promotion, monitoring and management of quality of goods, articles, processes, systems and services and to protect the interests of consumers and various other stakeholders and notify any other goods, articles, processes, systems and services for the purpose of sub-section (1) of section 16.

26. Restriction on use of name of Bureau and Indian Standard

(1) No person shall, with a view to deceive or likely to deceive the public, use without the previous permission of the Bureau—

(a) any name which so nearly resembles the name of the Bureau as to deceive or likely to deceive the public or the name which contains the expression "Indian Standard" or any abbreviation thereof; or

(b) any title of any patent or mark or trade mark or design, in relation to any goods, article, process, system or service, containing the expressions "Indian Standard" or "Indian Standard Specification" or any abbreviation of such expressions.

(2) Notwithstanding anything contained in any law for the time being in force, no registering authority shall—

(a) register any company, firm or other body of persons which bears any name or mark; or

(b) register a trade mark or design which bears any name or mark; or

(c) grant a patent, in respect of an invention, which bears a title containing any name or mark, if the use of such name or mark is in contravention of sub-section (1).

(3) If any question arises before a registering authority whether the use of any name or mark is in contravention of sub-section (1), the registering authority may refer the question to the Central Government whose decision thereon shall be final.

27. Appointment and powers of certification officers

（1）The Bureau may appoint as many certification officers as may be necessary for the purpose of inspection whether any goods, article, process, system or service in relation to which the Standard Mark has been used conforms to the relevant standard or whether the Standard Mark has been properly used in relation to any goods, article, process, system or service with or without licence, and for performing such other functions as may be assigned to them.

（2）Subject to any rules made under this Act, a certification officer shall have power to—

（a）inspect any operation carried on in connection with any goods, article, process, system or service in relation to which the Standard Mark has been used; and

（b）take samples of any goods or article or of any material or substance used in any goods, article, process, system or service, in relation to which the Standard Mark has been used.

（3）Every certification officer shall be furnished by the Bureau with a certificate of appointment as a certification officer, and the certificate shall, on demand, be produced by the certification officer.

（4）Every certified body or licence holder shall—

（a）provide reasonable facilities to certification officer to enable him to discharge the duties imposed on him;

（b）inform certification officer or the Bureau of any change in the conditions which were declared or verified by the certification officer or the Bureau at the time of grant of certificate of conformity or licence.

（5）Any information obtained by a certification officer or the Bureau from any statement made or information supplied or any evidence given or from inspection made under the provisions of this Act shall be treated as confidential:

Provided that nothing shall apply to the disclosure of any information for the purpose of prosecution and protection of interest of consumers.

28. Power to search and seizure

(1) If the certification officer has reason to believe that any goods or articles, process, system or service in relation to which the contravention of section 11 or sub-sections (6) or (8) of section 14 or section 15 or section 17 has taken place are secreted in any place, premises or vehicle, he may enter into and search such place, premises or vehicle for such goods or articles, process, system or service, as the case may be.

(2) Where, as a result of any search made under sub-section (1), any goods or article, process, system or service has been found in relation to which contravention of section 11 or sub-sections (6) or (8) of section 14 or section 15 or section 17 has taken place, the certification officer may seize such goods or article and other material and documents which, in his opinion will be useful for, or relevant to any proceeding under this Act:

Provided that where it is not practicable to seize any such goods or article or material or document, the certification officer may serve on the owner an order that he shall not remove, part with, or otherwise deal with, the goods or article or material or document except with the previous permission of the certification officer.

(3) The provision of the Code of Criminal Procedure, 1973 (2 of 1974), relating to searches and seizures shall, so far as may be, apply to every search or seizure made under this section.

29. Penalty for contravention

(1) Any person who contravenes the provisions of section 11 or sub-section (1) of section 26 shall be punishable with fine which may extend to five lakh rupees.

(2) Any person who contravenes the provisions of sub-sections (6) or (8) of section 14 or section 15 shall be punishable with imprisonment for a term which may extend to one year or with fine which shall not be less than one lakh rupees, but may extend up to five times the value of goods or articles produced or sold or offered to be sold or affixed or applied with a Standard Mark including Hallmark, or with both:

Provided that where the value of goods or articles produced or sold or offered to be sold cannot be determined, it shall be presumed that one year's production was in such contravention and the annual turnover in the previous financial year shall be taken as the value of goods or articles for such contravention.

(3) Any person who contravenes the provisions of section 17 shall be punishable with imprisonment for a term which may extend up to two years or with fine which shall not be less than two lakh rupees for the first contravention and not be less than five lakh rupees for the second and subsequent contraventions, but may extend up to ten times the value of goods or articles produced or sold or offered to be sold or affixed or applied with a Standard Mark, including Hallmark, or with both:

Provided that where the value of goods or articles produced or sold or offered to be sold cannot be determined, it shall be presumed that one year's production was in such contravention and the annual turnover in the previous financial year shall be taken as the value of goods or articles for such contravention.

(4) The offence under sub-section (3) shall be cognizable.

30. Offences by companies

Where an offence under this Act has been committed by a company, every director, manager, secretary or other officer of the company who, at the time the offence was committed, was in charge of and was responsible to the company for the conduct of the business of the company, or authorised representative of the company as well as the company, shall be deemed to be guilty of the offence and shall be liable to be proceeded against and punished accordingly, irrespective of the fact that the offence has been committed with or without the consent or connivance of, or is attributable to any neglect on the part of any director, manager, secretary or other officer of the company, or authorised representative of the company.

Explanation. —For the purposes of this section—

(a) "company" means a body corporate and includes a firm or other association of individuals;

(b) "director", in relation to a firm, means a partner in the firm.

31. Compensation for non-conforming goods

Where a holder of licence or certificate of conformity or his representative has sold any goods, article, process, system or service, which bears a Standard Mark not conforming to the relevant standard, or with colourable imitation, the certified body or licence holder or his representative shall be liable to compensate the consumer for the injury caused by such non-conforming goods, article, process, system or service in such manner as may be prescribed.

32. Cognizance of offence by courts

(1) No court inferior to that of a Metropolitan Magistrate or a Judicial Magistrate of the first class, specially empowered in this behalf, shall try any offence punishable under this Act.

(2) No court shall take cognizance of any offence punishable under this Act save on a complaint made by—

(a) or under the authority of the Bureau; or

(b) any police officer, not below the rank of deputy superintendent of police or equivalent; or

(c) any authority notified under sub-section (2) of section 16; or

(d) any officer empowered under the authority of the Government; or

(e) any consumer; or

(f) any association.

(3) Any police officer not below the rank of deputy superintendent of police or equivalent, may, if he is satisfied that any of the offences referred to in sub-section (3) of section 29 has been, is being, or is likely to be, committed, search and seize without warrant, the goods, die, block, machine, plate, other instruments or things involved in committing the offence, wherever found, and all the articles so seized shall, as soon as practicable, be produced before a Magistrate as prescribed under sub-section (1).

(4) The court may direct that any property in respect of which the contravention has taken place shall be forfeited by the Bureau.

(5) The court may direct that any fine, in whole or any part thereof, payable under the provisions of this Act, shall be payable to the Bureau.

33. Compounding of offence

(1) Notwithstanding anything contained in the Code of Criminal Procedure, 1973 (2 of 1974), any offence committed for the first time, punishable under this Act, not being an offence punishable with imprisonment only, or with imprisonment and also with fine, either before or after the institution of any prosecution, may be compounded by an officer so authorised by the Director General, in such manner as may be prescribed:

Provided that the sum so specified shall not in any case exceed the maximum amount of the fine which may be imposed under section 29 for the offence so compounded; and any second or subsequent offence committed after the expiry of a period of three years from the date on which the offence was previously compounded shall be deemed to be an offence committed for the first time.

(2) Every officer referred to in sub-section (1) shall exercise the powers to compound an offence, subject to the direction, control and supervision of the Bureau.

(3) Every application for the compounding of an offence shall be made in such manner as may be prescribed.

(4) Where any offence is compounded before the institution of any prosecution, no prosecution shall be instituted in relation to such offence against the offender in relation to whom the offence is so compounded.

(5) Where the composition of any offence is made after the institution of any prosecution, such composition shall be brought to the notice of the court in which the prosecution is pending in writing by the officer referred to in sub-section (1), and on such notice of the composition of the offence being given and its acceptance by the court, the person against whom the offence is so compounded shall be discharged.

34. Appeal

(1) Any person aggrieved by an order made under section 13 or sub-section (4) of section 14 or section 17 of this Act may prefer an appeal to Director General of the

Bureau within such period as prescribed.

(2) No appeal shall be admitted if it is preferred after the expiry of the period prescribed therefor:

Provided that an appeal may be admitted after the expiry of the period prescribed therefor if the appellant satisfies the Director General that he had sufficient cause for not preferring the appeal within the prescribed period.

(3) Every appeal made under this section shall be made in such form and shall be accompanied by a copy of the order appealed against and by such fees as may be prescribed.

(4) The procedure for disposing of an appeal shall be such as may be prescribed:

Provided that before disposing of an appeal, the appellant shall be given a reasonable opportunity of being heard.

(5) The Director General may suo motu or on an application made in the manner prescribed review the order passed by any officer to whom the power has been delegated by him.

(6) Any person aggrieved by an order made under sub-section (1) or sub-section (5) may prefer an appeal to the Central Government having administrative control of the Bureau within such period as may be prescribed.

35. Members, officers and employees of Bureau to be public servants

All members, officers and other employees of the Bureau shall be deemed, when acting or purporting to act in pursuance of any of the provisions of this Act, to be public servants within the meaning of section 21 of the Indian Penal Code (45 of 1860).

36. Protection of action taken in good faith

No suit, prosecution or other legal proceeding shall lie against the Government or any officer of the Government or any member, officer or other employee of the Bureau for anything which is in done or intended to be done in good faith under this Act or the rules or regulations made thereunder.

37. Authentication of orders and other instruments of Bureau

All orders and decisions of, and all other instruments issued by, the Bureau shall be authenticated by the signature of such officer or officers as may be authorised by the Bureau in this behalf.

38. Power to make rules

The Central Government may, by notification in the Official Gazette, make rules for carrying out the purposes of this Act.

39. Power to make regulations

The Executive Committee may, with the previous approval of the Central Government, by notification in the Official Gazette, make regulations consistent with this Act and the rules to carry out the purposes of this Act.

40. Rules and regulations to be laid before Parliament

Every rule and every regulation made under this Act shall be laid, as soon as may be after it is made, before each House of Parliament, while it is in session, for a total period of thirty days which may be comprised in one session or in two or more successive sessions, and if, before the expiry of the session immediately following the session or the successive sessions aforesaid, both Houses agree in making any modification in the rule or regulation or both Houses agree that the rule or regulation should not be made, the rule or regulation shall thereafter have effect only in such modified form or be of no effect, as the case may be; so, however, that any such modification or annulment shall be without prejudice to the validity of anything previously done under that rule or regulation.

41. Act not to affect operation of certain Acts

Nothing in this Act shall affect the operation of the Agricultural Produce (Grading and Marking) Act, 1937 (1 of 1937), or the Drugs and Cosmetics Act, 1940 (23 of 1940), or any other law for the time being in force, which deals with any standardisation or quality control of any goods, article, process, system or service.

42. Power to remove difficulties

(1) If any difficulty arises in giving effect to the provisions of this Act, the

Central Government may, by order, published in the Official Gazette, make such provisions not inconsistent with the provisions of this Act as may appear to be necessary for removing the difficulty:

Provided that no order shall be made under this section after the expiry of two years from the commencement of this Act.

(2) Every order made under this section shall be laid, as soon as may be after it is made, before each House of Parliament.

43. Repeal and savings

(1) The Bureau of Indian Standards Act, 1986 (63 of 1986), is hereby repealed.

(2) Notwithstanding such repeal, anything done or any action taken or purported to have done or taken including any rule, regulation, notification, scheme, specification, Indian Standard, Standard Mark, inspection order or notice made, issued or adopted, or any appointment, or declaration made or any licence, permission, authorisation or exemption granted or any document or instrument executed or direction given or any proceedings taken or any penalty or fine imposed under the Act hereby repealed shall, insofar as it is not inconsistent with the provisions of this Act, be deemed to have been done or taken under the corresponding provisions of this Act.

(3) The mention of particular matters in sub-section (2) shall not be held to prejudice or affect the general application of section 6 of the General Clauses Act, 1897 (10 of 1897), with regard to the effect of repeal.

附录3　印度标准局(BIS)—国际标准化研究院(NITS)—国际标准化组织(ISO)谅解备忘录(MOU)

Memorandum of Understanding between

The Bureau of Indian Standards(BIS),

National Institute of Training for Standardization (NITS)

and

The International Organization for Standardization(ISO)

Recognizing the International Organization for Standardization (hereinafter referred to as ISO) as the specialized international institution for matters of standardization, having as its objectives the facilitation of international exchange of goods and services and the development of cooperation in the sphere of intellectual, scientific, technological and economic activities, including the transfer of technology and good business practices to developing countries through the promotion of International Standards.

Bureau of Indian Standards, the National Standards Body is a Statutory Organization under the Department of Consumer Affairs, Ministry of Consumer Affairs, Food and Public Distribution, Govt., of India and is engaged in the following activities:

◆ Formulation of Indian Standards;

◆ Certification (Product & Management Systems);

◆ Product testing;

◆ Enquiry Point for WTO;

◆ Training.

To meet the growing needs and expectations of the industry, NITS was set up under the age is of BIS in 1995. Since then NITS has been organising various types of training programmes for industry as indicated in the training calendar. These programmes are conducted by a team of more than 250 well experiences, qualified, trained and able officers of BIS. NITS has also been organising International Training Programmes for developing countries in Asia, Africa, Europe and Latin America.

Keeping in view its expanding activities, NITS started functioning from its new premises in Noida from May, 2003.

Based on the outcome of bilateral discussions, the Secretary-General of ISO and the Director General of BIS (NITS) agreed upon the following Memorandum of Understanding to govern cooperation between their two organisations and the coordination of their activities in all areas where their functions and activities are complementary and mutually supportive.

1   Purpose of the Memorandum of Understanding

1.1 The purpose of this Memorandum of Understanding is to establish a partnership between the Parties, to contribute to training on standardization and related matters, in coordination with the respective ISO Regional Liaison Officers.

2   Fields of Cooperation

2.1 The Parties agree to cooperate, in the context of their respective mandates, policies and resources, for the purposes of capacity-building, training, including e-learning, in specific programmes identified by and agreed between both Parties.

2.2 The Parties will undertake joint projects and organize seminars and workshops at the regional level on standardization and related matters including Management Systems and Conformity Assessment Standards and Guides, within the following areas:

2. 2. 1 Joint organization of regional training events

2. 2. 2 Delivery of ISO modules within NITS training programmes

2. 2. 3 Provision of Need-Based Training in other Developing Countries

2. 2. 4 As part of this MOU, the ISO Technical Assistance and Training Services, DEVT, ISO Central Secretariat, will make available to NITS, training material and related documentation for the organization of training events foreseen in this MOU.

BIS acknowledges that these materials are protected by copyright and are only made available for use in connection with the training events agreed by the parties on condition that NITS preserves the integrity of such materials and their source is acknowledged.

Nothing in this MOU shall give NITS any right, title or interest in or to the Copyrights or to the ISO mark and logo save as granted hereby.

By mutual agreement, NITS may also develop training material for new training courses, which will be used after approval of ISO.

2. 3 The focal points for the purpose of this MOU shall be:

2. 3. 1 On behalf of BIS

NITS for technical content and management of progress and faculty identification and development.

IR&TISD for administrative and financial aspects.

2. 3. 2 On behalf of ISO

ISO Technical Assistance and Training Services, DEVT at the ISO Central Secretariat.

3　　Promotion, Exchange of Information

3. 1 ISO and NITS shall promote the training activities.

3. 2 ISO and NITS shall keep each other informed regularly about work of mutual interest carried out by each organization.

4   Administrative and Financial Modalities

4. 1 The implementation of the activities envisaged in this Memorandum of Understanding shall depend on the availability of the necessary financial resources, according to the regulations and procedures in force in the two organizations.

4. 2 In the execution of the agreed activities ISO and NITS shall support each other with the required logistic and administrative support.

4. 3 The modalities of joint projects under the present Memorandum of Understanding shall be contained in specific projects agreed upon by the Parties. The following focal points are specified for the purposes of this Memorandum of Understanding:

(a) For ISO: ISO Technical Assistance and Training Services, DEVT

(b) For BIS: As mentioned at 2. 3. 1

5   Final Clauses

5. 1 This Memorandum of Understanding may be amended by mutual written agreement of the Parties.

5. 2 Each party may terminate this agreement by giving the other party at least six months' written notice of such intention.

5. 3 Existing projects which are due to be completed within one calendar year from the date of notice of termination shall not be affected by such termination unless otherwise provided for.

5. 4 This Memorandum shall enter into force upon signature by the Secretary-General of ISO and the Secretary, Department of Consumer Affairs, Ministry of Consumer Affairs, Food and Public Distribution, Government of India and the Head of NITS.

For BIS   For ISO

Mr. L. Mansingh   Mr. Alan Bryden

Secretary, Department of   Secretary-General

Consumer Affairs

Ministry of Consumer Affairs

Food and Public Distribution

Government of India

For NTS

Mr. Anupam Kaul

Head

Date:_____

Place:_____

Date:_____

Place:_____

附录4 印度标准局(BIS)—印度工业联邦(CII)—美国国家标准协会(ANSI)建立印度—美国标准门户网站的谅解备忘录(MOU)

MEMORANDUM OF UNDERSTANDING (MOU)
TO ESTABLISH AN INDIA-U. S. STANDARDS PORTAL
BETWEEN
THE BUREAU OF INDIAN STANDARDS (BIS),
THE CONFEDERATION OF INDIAN INDUSTRY(CII)
AND
THE AMERICAN NATIONAL STANDARDS INSTITUTE(ANSI)

Recognizing the desirability of global cooperation in standards and conformity assessment, the Bureau of Indian Standards (hereinafter referred to as BIS), the Confederation of Indian Industry(hereinafter referred to as CII) and the American National Standards Institute (hereinafter referred to as ANSI) agree to facilitate the sharing of information on standards, conformity assessment, technical regulations (SCATR) and other trade related information in India and the United States, with the aim of facilitating understanding of the market access requirements in each market.

BIS, CII and ANSI agree to cooperate and establish Indian and U. S. content on a U. S. -India Standards Portal, hereinafter referred to as "the Portal".

The Portal will demonstrate a commitment by BIS, CII and ANSI to facilitate the sharing of information with the aim of enhancing Indo-U. S. trade.

Article 1

BIS, CII and ANSI agree to:

Set up a common Standards Portal for sharing the information related to:

● Standards

● Conformity Assessment

● Technical regulations

● Trade

Article 2

The Portal will consist of a single opening web page giving details of the objective of the Portal with brief overviews of BIS, CII & ANSI. The Portal will include hyperlinks to the websites of BIS, CII and ANSI, and to other relevant organizations in India and the U. S. , including standards development organizations (SDOs), bodies engaged in conformity assessment, technical regulations, government organizations, industry associations, and trade promotion organizations, ANSI will identify appropriate links to be included on the U. S. portion of the Portal, and BIS and CII will identify appropriate links to be included on the Indian side of the Portal.

Each organization shall additionally develop new resources in support of the Portal. These resources shall be stored separately on the respective organizations' servers, and shall be made available to Portal users through hyperlinks from the Portal. Maintenance and expansion of these resources shall be the responsibility of the organization that initially developed the resource.

The mechanism to make the Portal operational would be worked out jointly by BIS, CII and ANSI.

Article 3

BIS, CII and ANSI agree to:

1. Exchange views on the activity and operations of various programs of work related to standardization and conformity assessment in the U. S. and India as well as in regional and international organizations.

2. Encourage and support the cooperative engagement of technical experts (e. g. representatives industry consumer groups, government, etc.) from each other's economies. Such engagement could include formal or informal information sharing; working to develop mutually supportive positions in international fora; or organizing delegation visits, conferences, workshops and technical symposia on standardization and conformity assessment-related subjects of mutual interest.

3. Identifying and enabling communication between counterpart groups within the respective memberships and constituencies of BIS, ANSI and CII for the purpose of establishing sector or technology-specific relationships.

4. Work together on efforts to clarify and make easily available the market access requirements of each other's economies as they pertain to standards, conformity assessment (testing, certification, etc.) and technical regulations with the goal of facilitating Indo-U. S. trade.

5. Where appropriate, in all of the above, seek and share ideas for business opportunities for each other's members.

Article 4

Any issues arising from the interpretation and implementation of this MOU will be settled through mutual consultation among BIS, CII and ANSI or such other means as mutually decided.

Article 5

The agreement is effective from the date of signature.

The MOU may be amended by a written protocol for exchange of needs by all the three parties.

Any of the parties may terminate the MOU by providing the other parties a written notice at least six (6) months in advance.

Signed in New Delhi on the 20th December, 2007 in triplicate in the English language:

For the Bureau of	For the Confederation of	For the American National
Indian Standards (BIS)	Indian Industry(CII)	Standards Institute (ANSI)

Mr. Sayan Chatterjee	Lt. Gen S. S. Mehta (Retired)	Mr S. Joe Bhatia
Director Genera	Director General	President and CEO

附录5 印度标准局(BIS)—美国国家标准协会(ANSI)谅解备忘录(MOU)

MEMORANDUM OF UNDERSTANDING

BETWEEN

THE BUREAU OF INDIAN STANDARDS

AND

THE AMERICAN NATIONAL STANDARDS INSTITUTE

The Bureau of Indian Standards (BIS) and the American National Standards Institute (ANSI):

Recognizing the desirability of the global use of international standards;

Being committed to ensuring their national standards facilitate international trade, and pledging to facilitate cooperation between BIS and ANSI.

Hereby endorse the following articles of this MOU concerning matters covering standardization, conformity assessment and the mutual exchange of information and publications between the respective organizations.

Article 1

BIS and ANSI agree to:

(1) exchange views on the activity and operations and program of work of various internationaland regional standardization and conformity assessment organizations, including but not limited to the International Organization for Standardization (ISO) and the International Electrotechnical Commission (IEC);

(2) share each other's views and where feasible, work to develop mutually

supportive positions relating to the role of those organizations to their respective members;

(3) Exchange information regarding the system of accreditation of bodies involved in standardization and conformity assessment.

Article 2

BIS and ANSI agree to:

(1) facilitate dialogue among Indian and US experts in various areas of standardization, conformity assessment and quality assurance as determined by special arrangements between the parties;

(2) encourage standards organizations in their countries to identify their counterpart groups for the purpose of establishing bilateral relationships, including mutual exchange of information and materials, trade shows/exhibits to stimulate industry-to-industry cooperation as a building block to international standardization and free trade;

(3) arrange exchange of visits of the senior management of BIS and ANSI;

(4) explore the possibility of organizing and sponsoring conferences, workshops technical symposia on standardization and conformity assessment subjects of mutual interest.

Article 3

BIS and ANSI agree to exchange, using all appropriate media:

(1) their current catalogues to their national standards and procedures published by the respective organizations;

(2) general information and publications on conformity assessment published by the respective organizations;

(3) general materials for training courses on standardization, conformity assessment and quality assurance developed and published by the respective organizations;

(4) periodical publications and other relevant information published by the respective organization. Further to this item, BIS gave ANSI's permission to translate

and reprint free article in any and editions of the ANSI Reporter.

Article 4

Any issues arising from the interpretation or implementation of this MOU will be settled through consultations between BIS and ANSI or such other means as they may mutually decide.

Article 5

(1) This MOU may be amended by a written protocol or an exchange of notes of both Parties.

(2) Either Party may terminate this MOU by providing the other Party with a written notice at least six months in advance.

(3) Provisions of this MOU upon the expiry of its validity shall remain in force in relation to the projects being at the stage of their implementation.

Signed in Ottawa on the 13th September, 2006, in duplicate in the English language.

FOR THE BUREAU OF INDIAN STANDARDS
Secretary, Department of Consumer Affairs
Ministry of Consumer Affairs, Food Public Distribution
Mr. L. Mansingh

FOR THE AMERICAN NATIONAI, STANDARDS INSTITUTE
President and Chief Executive Officer

Mr. S Joc Bhatia

附录6　印度标准局(BIS)—美国电气与电子工程师协会(IEEE)谅解备忘录(MOU)

Memorandum of Understanding ("MOU") between Bureau of Indian Standards, India (BIS)

and

The Institute of Electrical and Electronics Engineers, Incorporated, USA (IEEE) on cooperation in standardization

The Bureau of Indian Standards and the Institute of Electrical and Electronics Engineer incorporated, hereinafter collectively referred to as the "Parties" and individually referred to as the "Party".

The Bureau of Indian Standards (BIS) came into existence on 1st April, 1987 through an Act of Parliament dated 26th November, 1986. BIS, a Body Corporate under the admin is tractive control of Ministry of Consumer Affairs, Food and Public Distribution, Govt. of India.

The Institute of Electrical and Electronics Engineers, Incorporated (IEEE) is the world's largest engineering society with members in over 160 countries and focuses on advancing the theory and practice of electrical, electronics, and computer engineering, computer science, and related technologies.

The IEEE Standards Association (IEEE-SA) is a global standardization body within IEEE comprised of individual and corporate members and other contributing technologists, who develop consensus-based, electro-technical and electronic information and communication technology standards affecting a broad market base. The IEEE-SA provides a standards program that serves the global requirements of industry, government, and the public. The IEEE-SA is the only body that can speak for IEEE in the area of standardization.

HAVE REACHED the following understanding:

1. Objectives

IEEE and BIS have the common objective to perform and promote, directly or indirectly, international standardization.

IEEE and BIS have a reciprocal interest in gaining knowledge about the activities of the other organization, which may facilitate cooperation on items of common interest. IEEE and BIS have noted the necessity of structuring and strengthening their relationship and fostering a closer cooperation.

Therefore, IEEE and BIS establish this MOU with the following goals:

(1) Encourage communication between the two organizations.

(2) Promote shared knowledge of the standards development activities of each party.

(3) Facilitate liaisons between each other's technical groups and other cooperation where possible.

2. Specific collaborative activities

(1) Both Parties agree to exchange information on their specific standards development activities on a quarterly basis. IEEE will provide a no-cost subscription to BIS of IEEE-SA's member-only *IEEE-SA News*, which includes information on IEEE standards activities, and other relevant newsletters.

(2) IEEE and BIS may exchange from time to time as they may agree, free of charge, other information in areas of mutual interest.

(3) A liaison will be appointed from IEEE and BIS. The liaisons will serve as the focal point for all interactions between the Parties. Liaisons shall be selected upon initiation of this MOU and contact information shared between the Parties.

(4) IEEE will provide on the IEEE-SA website (standards. ieee. org) an electronic link to the BIS website (www. bis. org. in) using BIS's Logo on the IEEE-SA website, in accordance with the terms of Section II. 7 below.

(5) BIS will provide on the BIS website an electronic link to the IEEE-SA website using IEEE's Logo on the BIS website, in accordance with the terms of Section II. 7 below.

(6) BIS and IEEE may identify and link leaders/stakeholders/technical experts with each other's technical committees and help promote the exchange of information on technical matters related to the formulation of standards.

(7) Each Party grants the other Party the right to use the Party's trademarks and Logos only for the specific and limited purposes expressly authorized by this MOU, provided that each Party has the right to review and approve in writing thirty (30) days in advance of such use by the other party. Upon expiration or termination of this MOU for any reason, the foregoing grant of rights shall terminate immediately. Both organizations recognize the importance of protecting intellectual property. Except as affirmatively provided for in this MOU, nothing in this MOU is intended to result in the transfer of either Party's intellectual property, including but not limited to any trademarks, copyrights or patents, to the other Party without the Parties entering into a written agreement documenting such transfer.

(8) This MOU is not a commitment of financial resources. Any resource commitment will be negotiated, documented, and committed separately between the Parties, and in advance of the actual expenditure. Except as otherwise agreed to and documented in writing, each Party will be responsible for their own costs related to its participation in this MOU.

3. Other objectives and activities

IEEE grants BIS the right to adopt IEEE standards (the "IEEE Standards") as specified In Annex A, which is incorporated herein by reference.

4. Term

This MOU shall come into effect from the date of its signature and shall remain in effect until terminated by one of the Parties in accordance with Section V below. IEEE and BIS agree to formally review this MOU every two (2) years to ensure the stated goals set forth in Section I of the MOU are being achieved.

5. Termination

This MOU may only be terminated as follows:

(1) Either Party may terminate this MOU upon ninety (90) days prior written notice;

(2) By written mutual agreement of IEEE and BIS;

(3) The insolvency of or the petition by or on behalf of IEEE or BIS for bankruptcy or reorganization under bankruptcy laws or any assignment for the benefit of creditors;

(4) By either Party, in the event of a material breach of this MOU by the other Party, if such breach is not cured within thirty (30) days after written notice of such breach.

Upon termination or expiration of this MOU, all the rights granted pursuant to this MOU shall cease immediately and the Parties shall cease from distributing any information received pursuant to this MOU. Parties shall destroy any material obtained from the other party that is identified as confidential, not for disclosure, or proprietary. BIS will immediately cease any further adoption, modification, translation, or sale of IEEE Standards, but may continue to use, without further modification, those already adopted and translated.

The following provisions of this MOU and Annex A shall survive termination: V, XI, XII, XIV, XV, XVI, A4.1(b)-(d), A4.2(d)-(g), A4.3, A4.4, A4.5(c)-(g).

6. Contents of MOU

The Parties hereto covenant and agree that this MOU constitutes the complete agreement between Parties, supersedes all prior agreements with respect to the subject hereof, and may not be amended or modified, except by written instrument signed by both the Parties hereto or by their duly authorized representatives.

7. Dispute settlement

Any difference or dispute that may arise in the application or interpretation or implementation, of this MOU shall be settled amicably through negotiations and consultations between the parties.

8. Severability

The terms and conditions of this MOU are severable. If any condition of this MOU is deemed to be illegal or unenforceable under any rule of law, all other terms shall remain inforce. Further, the term and condition which is held to be illegal or unenforceable shall remain in effect as far as possible and in accordance with the intention of the Parties.

9. Force majeure

Neither Party shall be responsible for any delay or failure in performance resulting fromacts entirely beyond its control.

10. Independent contractors

The relationship between the Parties shall be that of independent contractual Parties and nothing in this MOU shall be construed to constitute either Party as an employee, agent or member of the other Party. Without limiting the foregoing, neither Party shall have authority to act for or to bind the other Party in any way, to make representations or warranties or to execute agreements on behalf of the other Party, or to represent that it is in any way responsible for the acts or omissions of the other Party.

11. Third party beneficiaries

Nothing in this MOU, whether expressed or implied, is intended to confer any rights or remedies under or by reason of this MOU on any persons other than the Parties to this MOU and to their respective successors and assigns.

12. Counterparts

This MOU may be executed in one or more counterparts, each of which when so executed and delivered shall be deemed an original, but all of which together shall constitute one and the same instrument, and a signature page sent by facsimile or digital copy shall be deemed to be the equivalent of an original.

13. Assignment

Neither Party may assign this MOU, or any of its rights, obligations or duties hereunder, without the prior written consent of the other Party.

14. Indemnification

BIS shall indemnify, defend, and hold harmless IEEE and its officers, directors, agents and employees from and against all claims, losses, expenses, fees (including attorneys' and expert witnesses' fees), costs and judgments that may be asserted against IEEE that result from (a) any modification or translation of an IEEE Standard by BIS, or (b) any claims by any third parties which are based upon or are the result of any breach of the warranties contained in this MOU.

15. Confidentiality; publicity

The terms of this MOU are strictly confidential and BIS shall not disclose them to any third party, except (1) with express written consent of IEEE, (2) as necessary to effectuate the terms of the MOU, or (3) as necessary to comply with any applicable law or court order.

Before publishing or releasing any information regarding this MOU or any BIS Adopted Standard (as defined in Section A4.4 of Annex A), BIS will provide IEEE with at least ten (10) days' notice and the opportunity to: (1) review a copy of the planned public release or publication; (2) designate any information in the public release as not subject to disclosure. BIS will not disclose any material designated by IEEE as not subject to disclosure. This nondisclosure obligation, however, shall not apply to any knowledge or information which is now published, which subsequently becomes generally publicly known in the form in which it was obtained by BIS, other than as a direct or indirect result of the breach of this MOU, or which BIS is required to disclose by an applicable law or court order.

16. Notices

Any and all notices shall be in writing, sent by registered or certified mail, return receipt requested, or by air courier, addressed to the Parties at their respective addresses specified below, and are effective when mailed. Alternately, either a facsimile transmittal or an express mail transmittal with a confirmation of receipt shall be acceptable. Either Party, by notice, may specify a different address.

If to BIS: The Bureau of Indian Standards

9 Bahadur Shah Zafar Marg

Manak Bhavan

New Delhi-110002

India

Attention: Director General

Tel: 911123237991, Fax:911123239399

If to IEEE: The Institute of Electrical and Electronics Engineers, Incorporated

Standards Activities

445 Hoes Lane, Piscataway, NJ 08854

the United States

Attention: Dr. Konstantinos Karachalios, Managing Director

Tel:(732)562-3820,Fax:(732)562-1571

Dated:	Dated:
Signature	Signature
Bruce Kraemer	M. J. Joseph
President, IEEE-Standards Association	Director General
On behalf of	
The Institute of Electrical and	BUREAU OF INDIAN STANDARDS
Electronics Engineers, Incorporated	(BIS)
(IEEE)	
445 Hoes Lane	9 Bahadur Shah Zafar Marg
Piscataway, NJ 08854	Manak Bhavan
US	New Delhi 110002. India
http://standards. ieee. org	http://www. bis. org. in

Annex-Adopt on Agreement

A1. OBJECTIVES

1. Leverage TEEE's standard portfolio to strengthen India's national standards system;

2. Promote greater input and content by India into IEEE Standards through greater country participation in IEEE's standards development process;

3. Promote the BIS acceptance and use of IEEE Standards.

A2. IEEE UNDERTAKINGS

IEEE agrees to the following:

To provide annually access to a full collection of IEEE Standards electronically at no charge for the sole purpose of identifying IEEE Standards for national adoption by BIS (in accordance with Section A4 below).

A3. BIS UNDERTAKINGS

BIS agrees to the following:

1. To promote IEEE Standards where desirable and appropriate.

2. To send an annual report to the IEEE-SA on its IEEE-SA-related activities and a list of the total number of adopted and translated Standards sold. To the extent possible, BIS will identify the adopted and translated Standards sold, and the Standards sold to various categories of purchasers (e. g. individuals, academia, government, corporation, etc.) according to the template attached hereto as Annex B.

A4. ADOPTION AND TRANSLATION OF IEEE STANDARD AS INDIAN NATIONAL STANDARDS

Subject to the terms and conditions below, and only for the term of this MOU, BIS may designate, adopt and translate existing IEEE Standards as Indian National Standards, as follows:

1. Adoption With No Changes

(a) IEEE grants BIS a license to adopt IEEE Standards without making any changes in content other than the identifying number (BIS Standards Adopted Without Change). BIS shall provide IEEE with 10 days notice to mourequests@ieee. org of (i) the IEEE Standard BIS plans to adopt without change by identifying the IEEE Standard number, title and year; and (ii) the BIS identifying number that will be used in connection with such standard.

IEEE shall provide to BIS the electronic file for the IEEE Standards BIS intends to adopt in accordance with this section, in order to prepare the Indian National Standard for publication. BIS shall provide the final electronic versions of all IEEE Standards it adopts to mourequests@ieee.org at least 10 days before these adoptions are made available in India via the BIS Web site or other means.

(b) IEEE shall retain its worldwide copyright ownership rights in the IEEE Standards and the BIS Standards Adopted Without Change. When distributing BIS Standards Adopted Without Change, BIS shall retain IEEE's copyright footer that appears on each page of approved IEEE Standards as well as in the front matter.

(c) IEEE grants BIS a non-exclusive, royalty-free license to publish, sell, advertise and distribute the BIS Standards Adopted Without Change.

(d) Nothing in this agreement shall limit IEEE's worldwide ability, including in the Republic of India, to reproduce, publish, distribute, use, sell, or engage in any other activity, in any format either now known or hereinafter created, regarding IEEE Standards. For purposes of clarity, should IEEE engage in any of these activities this will not create any financial obligation from IEEE to BIS.

2. Adoption With Country-specific Changes

(a) Except as provided in Sections A4.2, A4.3 and A4.5, the base IEEE Standard cannot be changed.

(b) BIS may modify an IEEE Standard only to the extent necessary to adapt the standard to the technical, natural, or regulatory environment of the Republic of India (Country-Specific Changes). Such Country-specific Changes shall be documented appropriately in a National Foreword and through the addition of front matter and/or informative or normative annexes. No Country-specific Changes may be made to the body of an IEEE Standard.

(c) IEEE grants BIS a license to adopt IEEE Standards as modified by Country

Specific Changes ("IEEE Standards Adopted With Country-specific Changes"). If BIS determines that Country-Specific Changes are required, BIS shall provide IEEE notice, within 10 days of such determination, to mourequests@ieee.org of (i) the IEEE Standard BIS plans to adopt with Country-Specific Changes by identifying the IEEE Standard number, title and year; (ii) the BIS identifying number that will be used in connection with such standard and further provide; (iii) the content of the new material comprising the Country-Specific Changes that will be included in the separate informative/normative annex if IEEE determines, in its sole discretion, that a County Specific Change is useful to be included in the normative part of the IEEE standard, this modification will be evaluated according to Section A4.3 ("Modifications to IEEE Standards") below. IEEE shall provide to BIS the electronic file for the IEEE Standards BIS intends to adopt in accordance with this section, in order to prepare the Indian National Standard for publication. BIS shall provide the final electronic versions of the adopted standards, including the Country-Specific Changes and any annexes to mourequests@ieee.org at least 10 days before these adoptions are made available in India via the BIS Web site or other means.

(d) To the extent possible under copyright law, and as between IEEE and BIS. BIS shall own the copyright of any country-specific changes as derivative works based on IEEE Standards. IEEE shall retain its worldwide copyright ownership rights of the base IEEE Standard adopted by BIS with any Country-Specific Changes. When distributing BIS Standards Adopted With Country-Specific Changes, BIS shall retain IEEE's copyright footer that appears on each page of approved IEEE Standards as well as in the front matter.

(e) IEEE grants BIS a non-exclusive, royalty-free license to publish, sell, advertiseand distribute the IEEE Standards Adopted With Country-Specific Changes prepared by BIS

(f) All of BIS's interest in IEEE Standards, including but not limited to use, sale, publication, and distribution, both with and without Country-Specific Changes, may not be assigned, licensed, or transferred to any third party without the express, written consent of IEEE.

(g) BIS acknowledges that any license or representation obtained by IEEE from third parties regarding any patents or other rights applicable to an IEEE Standard may not extend to any modified standard, including any IEEE Standards Adopted With Country-Specific Changes. IEEE makes no representation regarding the applicability of any such license or representation to any IEEE Standard modified by BIS.

3. Modifications to IEEE Standards

(a) IEEE Standards are developed wholly by IEEE (except in the case of jointly developed standards). IEEE Standards have undergone intense technical scrutiny and in many cases, have been widely implemented in global markets by the time they are adopted by national bodies around the globe. Any modifications to these particular documents will affect the integrity of the IEEE Standard; hence, any suggested changes to published IEEE Standards must be submitted to and approved by IEEE in accordance with IEEE procedures.

(b) If BIS is interested in proposing changes to an existing IEEE Standard that are not exclusively Country-Specific Changes as detailed in Section A42 or Translations as detailed in Section A4. 5, BIS shall contact the IEEE Sponsor for that standard to determine the feasibility of revising the standard and engaging in such activity.

(c) All changes are subject to approval by IEEE's established consensus-based ballot. IEEE does not guarantee in advance the outcome of any such ballots.

All contributions to IEEE Standards are subject to the IEEE-SA Intellectual Property Rights Policies.

4. Notice Statements on BIS Adopted Standards

BIS agrees to place the following statement on any BIS Standards Adopted With No Changes and BIS Standards Adopted With Country-Specific Changes (collectively, "BIS Adopted Standards"):

"This Indian National Standard is a licensed product of IEEE and based on IEEE XXX-XX, (Title, Year, Date), 445 Hoes Lane Piscataway, NJ, 08854, USA. This

Indian National Standards is only valid in India as an Indian National Standard. No changes have been made to the base IEEE Standard except those that may be included as Country-Specific Changes in a front matter or an attached informative annex."

5. Translations

(a) IEEE grants BIS a license to translate BIS Adopted Standards developed in compliance with this Annex A into Hindi (BIS Translations) maintaining as much as possible the integrity of the IEEE Standards, in accordance with the terms and conditions of this section.

(b) BIS shall provide IEEE with 10 days notice to mourequests@ieee. org of the BIS Adopted Standard BIS plans to translate by identifying the number, title and year.

(c) BIS will place the following IEEE translation notice on any BIS Translation (in both English and Hindi) in a prominent location in the translated document:

"IEEE has authorized the translation of this standard by BIS, which is responsible for the technical and linguistic accuracy of the translation. Only the English edition as published and copyrighted by IEEE shall be considered the official IEEE version."

(d) BIS will place the following statement on any BIS Translation (in both English and Hindi) in a prominent location in the translated document:

"This Indian Standard is based on IEEE XXX-XX, (Title, Year, Date) Translated and reprinted pursuant to license agreement with IEEE."

(e) BIS shall provide the final electronic versions of all IEEE Standards it adopts and translates to mourequests @ieee. org at least 10 days before these adoptions are made available in India via the BIS Web site or other means.

(f) IEEE shall own the copyright in the existing IEEE Standards as well as the copyright of the translated Hindi documents.

(g) IEEE grants BIS a non-exclusive, royalty-free license to publish, sell, advertise and distribute the BIS Translations.

A5. GENERAL TERMS AND CONDITIONS

1. Enforcement

(a) Should BIS become cognizant of any violation of IEEE's intellectual property rights, BIS shall provide notice to IEEE of any infringements in India of the BIS Adopted Standards or BIS Translations. The Parties shall consult with each other to develop an enforcement plan with respect to such infringements. Absent any agreement with regard to an enforcement plan, the party or Parties who own copyright rights within the territory of India may, within their discretion, take such actions as may be deemed appropriate to protect and enforce their copyright.

(b) IEEE grants BIS a non-exclusive, non-transferrable right to enforce the copyright of BIS Adopted Standards in India. The option to terminate this license shall reside exclusively with IEEE and may be invoked immediately upon written notice to BIS. Termination of this license is severable to MOU. BIS agrees that it shall not commence litigation, or any event in anticipation of litigation, with respect to the BIS Adopted Standards without the express, written consent of IEEE. BIS shall not enforce any IEEE intellectual property except as permitted under this MOU.

2. BIS may place its logo on the BIS Adopted Standards and BIS Translations and assign a BIS number to such standards according to the rules and regulations of the Bureau of Indian Standards.

3. BIS may create and administer programs to certify compliance with BIS Adopted Standards (Certification Programs). BIS shall inform IEEE upon the creation of each new Certification Program and shall report to IEEE, no less frequently than once per year, the entities and products that have received certification under any Certification Program.

4. Except as affirmatively permitted above, BIS shall not create any derivative works based on IEEE Standard.

5. BIS shall not have the right to assign or sublicense to any third party or entity, in India or outside of India, any rights granted herein by IEEE with respect to the BIS Adopted Standards and BIS Translations.

6. BIS agrees not to submit or provide IEEE Standards, BIS Adopted Standards and/or BIS Translations to any other standards developers or organizations (whether

within India, national, international or other) for use, review or approval, without the prior written consent of the IEEE Standards Association Managing Director, except as required for circulation amongst technical committee members for the purpose for adoption of IEEE Standards according to BIS rules.

7. For the avoidance of doubt, IEEE shall retain copyright to its library of IEEE Standards.

8. The requirements of Section A4.2 regarding Country-Specific Changes shall apply to standards, adopted by BIS, that are promulgated by other standard development organizations (including ISO, IEC, ITU, and JTC1) but based on IEEE Standards.

ANNEX B- ADOPTION REPORT TEMPLATE

Sales/Downloads of IEEE Adoptions

Sales Date	Type of Customer (i.e. Govt, students, etc)	Customer	Customer Country	Std Title	Stds Number	Quantity

附录7　印度标准局(BIS)—欧洲标准化委员会(CEN)印度在全国范围内采用欧洲标准化委员会的 EN 115：1995 ＋ A1：1998 ＋ A2：2004"自动扶梯和乘客输送机的建造和安装安全规则"谅解备忘录(MOU)

Memorandum of Understanding(MOU)

for the use of the content of

EN 115：1995＋A1：1998＋A2：2004 for national adoption in India

By this MOU，THE EUROPEAN COMMITTEE FOR STANDARDIZATION，with registered office in Avenue Marnix 17，B-1000 Brussels，Belgium（hereinafter referred to as "CEN"）and THE BUREAU OF INDIAN STANDARDS，with registered offices in Manak Bhavan，9 Bahadur Shah Zafar Marg New Delhi 110 002，India（hereinafter referred to as "BIS"）

Agree on the terms and conditions for the use by BIS of the content of the **EN 115：1995＋A1：1998＋A2：2004 Safety rules for the construction and installation of escalators and passenger conveyors**（hereinafter referred to as the "EN"）for the purpose of the development and distribution of the three national standards：DOC ET 25(6607)，DOC ET 25 (6608)and DOC ET 25(6609)（hereinafter referred to as the "National Standards"），in full respect of CEN copyright protection rules.

SPECIFIC CONDITIONS governing the implementation of this MOU between

CEN and BIS are as follows：

● BIS shall ensure that the foreword of each of the National Standards includes an acknowledgement that the text contains extracts of the EN, reproduced with the permission of CEN.

● BIS shall notify CEN of the publication of the National Standards using the content of the EN provided under the terms of this MOU, by sending an electronic copy of the National Standards to the attention of the External Relations Director at the following e-mail address：ENadoptions@ cenceneleceu. edu.

● BIS shall not represent the National Standards developed under this MOU as having the status of European Standards in any of its advertisements or other promotional material or activity or in any of its communications with customers or the general public.

● There shall be no limitation on marketing and sale of the National Standards developed under this MOU within the Indian Territory. BIS shall not seek customers or undertake active promotions of the National Standards outside the Indian Territory. For the purposes of this MOU, passive advertisement or sale via a web-site shall not be considered as active promotion.

● The distribution of the National Standards developed under this MOU shall not be subject to any royal.

● BIS shall keep an accurate record of all sales of the National Standards developed under this MOU and shall render a statement in writing to CEN each quarter, using the template provided by CEN to this effect.

● BIS shall refrain from using any Trade Mark or Trade Name resembling the Trade Mark or Trade Name of CEN, as to be likely to cause confusion or deception.

● CEN and BIS acknowledge that the present MOU is established for a limited duration of two years from the date of its signature. It should be repealed by a Framework License MOU to be signed by CEN and BIS within this two-year timeframe.

● Amendments may be made to this Memorandum of Understanding upon mutual agreement between CEN and BIS.

● The Parties shall solve amicably any problem arising under or relating interpretation or implementation of this Memorandum of Understanding.

● This Memorandum of Understanding may be terminated at any time by either Party giving at least one months' prior notice to the other Party.

8. 1. IN WITNESS WHEREOF，this MOU in English and Hindi is made out in two original copies and is duly signed by authorized representatives of CEN and BIS. In case of any difference in interpretation or discrepancy between the language versions，the English version of the Agreement shall prevail.

Signed for and on behalf of BIS   Signed for and on behalf of CEN

Mrs. Alka Panda   Mrs. Elena SANTIAGO CID

Director General   Director General

Date：   Date：

参考文献

［1］ SAMPATH S. A tribute to Dr. Lal C. Verman［J］. IETE journal of research，1980，26(8)：i-ii.

［2］ Bureau of Indian Standards. Origin of BIS［EB/OL］.［2021-01-16］. http：//www. bis. org. in/bis_origin. asp.

［3］ 赵鸣歧. 印度之路：印度工业现代化道路探析［M］. 上海：学林出版社，2005.

［4］ Ministry of Law and Justice，Government of India. The Bureau of Indian Standards Act，1986［EB/OL］.（1986-12-23）［2021-01-16］. http：//lawmin. nic. in/ld/PACT/1986/The％ 20Bureau％ 20of％ 20Indian％ 20Standards％ 20Act，％ 201986. pdf.

［5］ Ministry of Food ＆ Civil Supplies. The Bureau of Indian Standards Rules，1987［EB/OL］.（1987-03-21）［2021-01-16］. http：//www. bis. org. in/bs/bisrules. htm.

［6］ Ministry of Health and Family Welfare，Government of India. The Drugs and Cosmetics Act，1940 (as amended up to the 30th June，2005)［EB/OL］.（2015-06-30）［2021-01-16］. http：//cdsco. nic. in/writereaddata/Drugs ＆ CosmeticAct. pdf.

［7］ Bureau of Indian Standards. National Building Code of India 2016［EB/OL］.［2021-01-16］. http：//bis. org. in/sf/nbc. asp.

［8］ Ministry of Law and Justice，Government of India. The Bureau of Indian Standards Act，2016［EB/OL］.（2016-03-21）［2021-01-22］. http：//www. indiacode. nic. in/acts-in-pdf/2016/201611. pdf.

［9］ Planning Commission，Government of India. Twelfth Five Year Plan（2012—2017）：Economic Sectors［EB/OL］.（2014-04-13）［2021-01-23］. http：//planningcommission. nic. in/plans/planrel/fiveyr/12th/pdf/12fyp_vol2. pdf.

[10] Press Information Bureau，Government of India. Meeting on strengthening regulatory and standards framework［EB/OL］.（2017-07-03）［2021-02-03］. http：//pib. nic. in/newsite/PrintRelease. aspx？relid＝167104.

[11] The Economic Times. Restrictions likely on import of items that hit local companies［EB/OL］.（2017-07-12）［2021-02-03］. https：//economictimes. indiatimes. com/news/economy/policy/restriction-likely-on-import-of-items-that-hit-local companies/articleshow/59552477. cms.

[12] Bureau of Indian Standards. Purposes & Objectives［EB/OL］.［2021-02-07］. http：//www. bis. org. in/org/obj. htm.

[13] WIEMANN J. Green protectionism：a threat to third world export？［M］// PIETER V D M，SIDERI S. Multilateralism versus regionalism：trade issues after the uruguay round. Taylor & Francis，2005：114.

[14] The Indian Express. New foreign trade policy：＄900 bn exports by FY20［EB/ OL］.（2015-04-02）［2021-02-12］. http：//indianexpress. com/article/business/ business-others/new-foreign-trade-policy-900-bn-exports-by-fy20/.

[15] The Hindu Business Line. Exporters must be informed of change in import norms in foreign nations quickly：sitharaman［EB/OL］.（2017-05-01）［2021-02- 12］. http：//www. thehindubusinessline. com/economy/policy/exporters-must-be-informed-of-change-in-import-norms-in-foreign-nations-quickly-sitharaman/article 9675576. ece.

[16] Confederation of Indian Industry. Nirmala sitharaman launches India standards portal［EB/OL］.（2017-05-01）［2021-02-15］. http：//www. cii. in/Pressreleases Detail. aspx？enc＝zay9Ym7sP/tsuraOqH4538DqkdLwXjWo＋51Vm/1eWH4＝.

[17] Press Information Bureau，Government of India. 4th National Standards Conclave［EB/OL］.（2017-04-28）［2021-02-15］. http：//pib. nic. in/newsite/ PrintRelease. aspx？relid＝161366.

[18] Dairy Reporter. Majority of Indian milk violates standards-government［EB/ OL］.（2012-10-22）［2021-02-16］. https：//www. dairyreporter. com/Article/2012/10/ 22/Majority-of-Indian-milk-violates-standards-government.

[19] Business Today. FSSAI launches comprehensive IT platform for uniform regulation of food standards [EB/OL]. (2017-11-02)[2021-02-16]. http://www. businesstoday. in/current/policy/fssai-launches-comprehensive-it-platform-for-uniform-regulation-of-food-standards/story/263203. html.

[20] Ministry of Electronics and Information Technology, Government of India. National Telecom Policy-2012[EB/OL]. [2021-02-25]. http://meity. gov. in/writereaddata/files/National％20Telecom％20Policy％20（2012）％20（480％20KB）. pdf.

[21] Livemint. Private standards have become strong: Rita Teaotia[EB/OL]. (2016-03-19)[2021-02-25]. http://www. livemint. com/Politics/hGfGfTBgntkY5Nwr XLMU1O/Private-standards-have-become-strong-Rita-Teaotia. html.

[22] FICCI. Lack of harmonized food standards affecting exports: FSSAI chairman [EB/OL]. (2013-03-15)[2021-02-25]. http://ficci. in/pressrelease/1176/ficci-press-release-food-standards. pdf.

[23] Food Safety Helpline. Food safety and standards authority of India calls for harmonisation of India's food standards with codex standards and other international best practices [EB/OL]. (2013-03-11)[2021-02-25]. http://foodsafetyhelpline. com/2013/03/fssai-harmonisation-with-codex/.

[24] National Accreditation Board for Hospitals & Healthcare Providers. Hospital accreditation: achievements & international linkages[EB/OL]. [2021-03-01]. http://www. nabh. co/Hospitals. aspx.

[25] Livemint. Narendra Modi launches smart city projects in pune[EB/OL]. (2016-06-25)[2021-03-01]. http://www. livemint. com/Politics/OBAdsQSOHayTH9 FoYeW9VL/Narendra-Modi-launches-smart-city-projects-in-Pune. html.

[26] Ministry of Power, Government of India. Notification[EB/OL]. (2014-01-30) [2021-03-02]. http://www. egazette. nic. in/WriteReadData/2014/158019. pdf.

[27] Ministry of Housing and Urban Affairs, Government of India. Smart city mission statement and guidelines[EB/OL]. [2021-03-02]. https://smartnet. niua. org/content/2dae72ca-e25b-4575-8302-93e8f93b6bf6.

［28］ Ministry of Housing and Urban Affairs，Government of India. Smart city mission dashboard［EB/OL］.［2021-03-02］. https：//smartnet. niua. org/smart-cities-network.

［29］ Business Standard. In a first，BIS to come up with standards for smart cities ［EB/OL］.（2015-06-05）［2021-03-05］. http：//www. business-standard. com/article/economy-policy/in-a-first-bis-to-come-up-with-standards-for-smart-cities-115060400931_1. html.

［30］ Bureau of Indian Standards. Draft Indian standard：smart cities-indicators［EB/OL］.（2016-09-30）［2021-03-05］. http：//www. bis. org. in/sf/ced/CED59（10000）_30092016. pdf.

［31］ Ministry of Communication and Information Technology. National Cyber Security Policy-2013（NCSP-2013）［EB/OL］.（2013-07-02）［2021-03-06］. http：//meity. gov. in/sites/upload_files/dit/files/National％20Cyber％20Security％20Policy％20％281％29. pdf.

［32］ The Economic Times. Government finalising cyber security standards for mobile phones［EB/OL］.（2017-09-01）［2021-03-06］. https：//economictimes. indiatimes. com/news/economy/policy/government-finalising-cyber-security-standards-for-mobile-phones/articleshow/60315930. cms.

［33］ First Pos. Government asks Chinese smartphone makers to outline security and privacy features used in devices［EB/OL］.（2017-08-16）［2021-03-06］. http：//www. firstpost. com/tech/news-analysis/government-asks-chinese-smartphone-makers-to-outline-security-and-privacy-features-used-in-devices-3937747. html.

［34］ Livemint. Govt planning to draft legal framework for cybersecurity standard［EB/OL］.（2017-08-25）［2021-03-10］. http：//www. livemint. com/Politics/xvFMv8slC2uJLnxkcUxIrO/Govt-planning-to-draft-legal-framework-for-cybersecurity-sta. html.

［35］ Bureau of Indian Standards. Green initiative by BIS-rooftop solar power plants at BIS offices［EB/OL］.［2021-03-10］. http：//bis. org. in/other/solar_power. asp.

［36］ Ministry of New and Renewable Energy，Government of India. Quality certification，standards and testing for grid-connected rooftop solar PV systems/

power plants [EB/OL]. [2021-03-11]. http://mnre. gov. in/file-manager/UserFiles/Rooftop-Solar-PV-Quality-Standards_Revised. pdf.

[37] The Economic Times. India issues new specifications for solar power modules，[EB/OL]. （2017-08-31）[2021-03-11]. https://energy. economictimes. indiatimes. com/news/renewable/india-issues-new-specifications-for-solar-power-modules/60310141.

[38] India Standards Portal. National Institute of Training for Standardization[EB/OL]. [2021-03-12]. http://indiastandardsportal. org/Detail. aspx? MenuId=14.

[39] National Institute of Training for Standardization. Training programs on standardization[EB/OL]. [2021-03-12]. http://10. 6. 0. 124/files/10380000005 37D5C/www. bis. org. in/trg/Prog_Standardization. pdf.

[40] Bureau of Indian Standards. Training calendar for 2016—2017，on-campus (open) programmes[EB/OL]. [2021-03-15]. http://www. bis. org. in/trg/Trg_Calender_16_17. pdf.

[41] National Institute of Training for Standardization. International training programmes for developing countries 2017—2018 [EB/OL]. [2021-03-15]. http://www. bis. org. in/trg/Int_Training. asp.

[42] Institute of Quality，Confederation of Indian Industry. Quality management systems[EB/OL]. [2021-03-15]. http://www. cii-iq. in/quality_management_systems.

[43] VIBHIR RELHAN. How effective is the current procedure of product quality standardization in the Indian market? A case study on Bureau of Indian Standards (BIS)- national standards body[J]. Centre for civil society，working paper No. 281，2012.

[44] Bureau of Indian Standards. Bureau of Indian Standards launches scheme for recognition of consumer organizations[EB/OL]. [2021-03-15]. http://bis. org. in/other/Consume_Recog. pdf.

[45] Bureau of Indian Standards. User manual for MANAK online for BIS[EB/OL]. [2021-03-15]. http://manakonline. in/MANAK/resources/app_srv/Manuals/

applicant_new. pdf.

[46] Confederation of Indian Industry. Standards conclave 2015 role of standards in international trade and use as non-tariff measures challenges and opportunities[EB/OL]. [2021-03-16]. http://www. cii. in/Digital_Library_Details. aspx? enc＝pZVQM37jt SRTHIkm BsithUf1rJfb4XdPpVIsKrBc95rfxT7OhsFqxKy2ZI＋zlpUH.

[47] Press Information Bureau, Government of India. 4th National Standards Conclave[EB/OL]. (2017-04-28)[2021-03-16]. http://pib. nic. in/newsite/ PrintRelease. aspx? relid＝161366.

[48] International Organization for Standardization. TC participation–BIS India[EB/OL]. [2021-03-20]. https://www. iso. org/member/1794. html? view＝participation&t＝S.

[49] IBSA. Action plan on trade facilitation for standards, technical regulations and conformity assessment[EB/OL]. (2006-09-13)[2021-03-20]. http://www. ibsa-trilateral. org/images/stories/documents/agreements/20060913IBSAActionPlan. pdf.

[50] IBSA. Memorandum of understanding on trade facilitation for standards, technical regulations and conformity assessment among the government of the Republic of India, the government of the Federative Republic of Brazil and the government of the Republic of South Africa[EB/OL]. (2008-10-15)[2021-03-20]. http://ibsa. nic. in/mou_t&i. htm.

[51] Statistics Times. Sector-wise contribution of GDP of India[EB/OL][2021-03-20]. http://statisticstimes. com/economy/sectorwise-gdp-contribution-of-india. php.

[52] The Hindu Business Line. Standardisation in services sector to help boost trade: official[EB/OL]. (2015-05-04)[2021-03-22]. http://www. thehindubusinessline. com/economy/standardisation-in-services-sector-to-help-boost-trade-official/article 7170372. ece.

[53] JAIN S. Regional cooperation in South Asia: India perspectives[M]//AHMED S, KELEGAMA S,GHANI E. Promoting economic cooperation in South Asia: Beyond SAFTA. SAGE Publications, 2010:300-320.

[54] American National Standards Institute. Memorandum of Understanding Between

the Bureau of Indian Standards and the American National Standards Institute ［EB/OL］. （2006-09-13） ［2021-03-22］. https：//share. ansi. org/Shared％ 20Documents/About％ 20ANSI/Memoranda％ 20of％ 20Understanding/ANSI-BIS％ 20MOU％20（Signed％20-％20Bureau％20of％20Indian％20Standards）. pdf.

［55］ Bureau of Indian Standards. BIS, CII & ANSI （USA） signs MOU on establishment of India－US standards portal［EB/OL］. ［2021-03-24］. http：// www. bis. org. in/ansimouwriteup. pdf.

［56］ ANSI Standards Portal. Exporting to India：key information［EB/OL］. ［2021-03-24］. https：//www. standardsportal. org/usa_in/key_information. aspx.

［57］ Bureau of Indian Standards. Bureau members［EB/OL］. ［2021-03-26］. http：// www. bis. org. in/org/compbureau. asp.

［58］ Bureau of Indian Standards. Composition of textile division council［EB/OL］. ［2021-03-26］. http：//www. bis. org. in/sf/comptxd. pdf.

［59］ Bureau of Indian Standards. Manual for Standards Formulation （first revision） ［EB/OL］. ［2021-04-01］. http：//www. bis. org. in/qazwsx/TD/manual. pdf.

［60］ Bureau of Indian Standards. Mandatory certification［EB/OL］. ［2021-04-01］. http：//www. bis. org. in/cert/ProdUnManCert. asp.

［61］ Food Safety and Standards Authority of India. FSSAI Adopts Vertical Standards for food additive［EB/OL］. ［2021-04-01］. https：//gain. fas. usda. gov/Recent％ 20GAIN％ 20Publications/FSSAI％ 20Adopts％ 20Vertical％ 20Standards％ 20for％20Food％20Additives_New％20Delhi_India_1-5-2017. pdf.

［62］ Bureau of Indian Standards. Technical Committees：food and agriculture［EB/ OL］. ［2021-04-01］. http：//164. 100. 105. 199：8071/php/BIS/TechnicalSub Committees. php？Name＝Food％20and％20agriculture；Food Safety and Standards Authority of India，"List of E Mail IDs for Scientific Committee and Panels，" http：//www. fssai. gov. in/home/food-standards/scientific-committee. html.

［63］ Food Safety Helpline. FSSAI addresses public concerns on packaged drinking water［EB/OL］. （2016-07-12） ［2021-04-01］. http：//foodsafetyhelpline. com/ 2016/07/fssai-addresses-public-concerns-packaged-drinking-water/.

［64］ The Economic Times. FSSAI wants action against unlicensed water packaging units［EB/OL］. (2016-06-28)［2021-04-02］. https://economictimes. indiatimes. com/industry/cons-products/food/fssai-wants-action-against-unlicensed-water-packaging-units/articleshow/52957903. cms.

［65］ BHARDWAJ R. Standard setting in India: competition law and IP issues［J］. Indore management journal，2013，5:92-101.

［66］ Export Inspection Council of India. Export Inspection Council of India contents ［EB/OL］. ［2021-04-02］. http://www. eicindia. gov. in/Services/Compliance/ Preferential-Tariff-Schemes. pdf.

［67］ Export Inspection Council of India. Health: Guidelines for Issuance of Health Certificate for Fishery Products Meant for Export to Countries of the European Union［EB/OL］. ［2021-04-02］. http://www. eicindia. gov. in/Services/Compliance/ Health-Certificate. aspx.

［68］ Export Inspection Council of India. Issue of certificate of authenticity for European Union［EB/OL］. ［2021-04-02］. http://www. eicindia. gov. in/Services/ Compliance/authenticity. pdf.

［69］ Export Inspection Council. International recognitions［EB/OL］.［2021-04-03］. http:// www. eicindia. gov. in/About-EIC/About-US/International-Recognitions. aspx.

［70］ Bureau of Indian Standards. Product certification［EB/OL］. ［2021-04-03］. http://www. bis. org. in/home_product. asp.

［71］Bureau of Indian Standards. Ministry of Steel Order［EB/OL］. (2012-03-12) ［2021-04-03］. http://bis. org. in/MandatoryProducts/QCOrder/Steel％20QCO/ SO％20No. ％20414％20(E). pdf.

［72］ Bureau of Energy Efficiency，Government of India. Standards & labeling［EB/ OL］. ［2021-04-05］. https://beeindia. gov. in/content/standards-labeling.

［73］ International Telecommunication Union. Member states list: India［EB/OL］. ［2021-04-05］. https://www. itu. int/online/mm/scripts/gensel9? _ ctryid ＝ 1000100560&_ctryname＝India.

［74］ KNN Knowledge & News Network. Stainless-steel MSMES hit hard as mandatory BIS certification maintained [EB/OL]. (2017-04-18) [2021-04-05]. http://knnindia. co. in/ news/newsdetails/msme/stainless-steel-msmes-hit-hard-as-mandatory-bis-certification-maintained.

[75] International Trade Administrations, U. S. India - standards for trade[EB/OL]. (2017-02-08)[2021-04-06]. https://www. export. gov/article? id = India-Trade-Standards.

[76] NITI Aayog, Government of India. Make in India strategy for electronic products [EB/OL]. [2021-04-06]. http://niti. gov. in/writereaddata/files/document_publication/ Electronics%20Policy%20Final%20Circulation. pdf.

[77] Ministry of Commerce and Industry, Government of India. Export import data bank (annual)[EB/OL]. [2021-04-10]. http://commerce. gov. in/EIDB. aspx.

[78] Richa Sekhani. Tackling India's trade deficit with China[EB/OL]. (2015-04-23)[2021-04-10]. http://www. futuredirections. org. au/publication/tackling-india-s-trade-deficit-with-china.

[79] Reuters. India's tightened consumer goods standards could hurt China imports [EB/OL]. (2017-10-18)[2021-04-10]. https://www. reuters. com/article/india-china-trade/indias-tightened-consumer-goods-standards-could-hurt-china-imports-idUSL4N1MM23S.